HISTORICAL GEOGRAPHY

Prepared under the special
direction and scientific
Editorship of

PROFESSOR FRANK DEBENHAM
Cambridge University

Uniform with this book and completing the series in five volumes

THE USE OF GEOGRAPHY
F. Debenham

ECONOMIC GEOGRAPHY
W. S. Thatcher

GEOGRAPHY OF LIVING THINGS
M. S. Anderson

PHYSICAL GEOGRAPHY
R. F. Peel

HISTORICAL GEOGRAPHY

J. B. MITCHELL, M.A.
Fellow of Newnham College, Cambridge
University Lecturer in Geography

THE ENGLISH UNIVERSITIES PRESS LTD
ST. PAUL'S HOUSE WARWICK LANE
LONDON, E.C.4

First printed 1954
Second Impression 1960
Third Impression 1963
Fourth Impression 1965
Fifth Impression 1967

SBN 340 05889 7

Printed and bound in Great Britain for the English Universities Press, Ltd., London, by C. Tinling & Co., Ltd., Liverpool, London and Prescot.

PREFACE

This book is a pioneer effort; it is an attempt to give readers outside the University Schools of Geography some idea of what Historical Geography is, and to interest them in it, so that they may find pleasure in pursuing its study for themselves. The plan adopted has been to work from the local to the wider scene, for study in the field is the key to all sound geographical work. Readers will find it helpful to use as they read an atlas of the British Isles, for example Bartholomew's Road Atlas (which also forms the atlas section of the A.A. Road Books) or the Oxford Travel Atlas of Britain. Suggestions have been made of exercises that may be carried out locally, not necessarily for the intrinsic value of the results, but in order to gain understanding of the kind of questions that must be asked and the nature of the data that must be collected to throw light upon the character of a place in the past, both as a distinctive region and as an integral part of the then known world. It is realised that much of the material suggested for use in independent study needs expert handling if the deductions drawn are to be sound in detail; but beginners must learn by trial and error.

Consideration of historical political geography has been omitted; any study of the areas of states, their frontiers and their capitals, most emphatically demands the historical approach. The topic requires a book to itself. It has been omitted the less reluctantly since, first, political geography is the one branch of geography that already is almost always studied in past periods as well as in the present and the historical approach is here a familiar one; and, secondly, the major lineaments of the political geography within Britain have long been simple and unchanging, so that the study of political geography, except in a very special sense, does not lend itself to local treatment which is the plan of this book.

This book would never have been written without the very great help that my colleagues have given in discussion and encouragement. I would like to thank particularly the Specialist editor of this series, Professor Debenham, and Mr. W. S. Thatcher. I am very grateful to Mr. G. Willett for help with the index, and to Mr. L. T. Thurston who has drawn all the diagrams for me, except Fig. 22 for which I thank Miss M. C. Anderson. Figs. 6, 9, 10, 12, 13, 19, 23, 24, 25, 29, 30, 32 and 33 are taken from Ordnance Survey maps and are reproduced here by permission of the Director General Ordnance Survey. Many other maps and diagrams have been reproduced by permission of the authors to whom acknowledgement is made in the text. I am much indebted to them. I thank also the editor-in-chief of the series Mr. Leonard Cutts for great patience in waiting for a manuscript long delayed.

7 *August*, 1953.

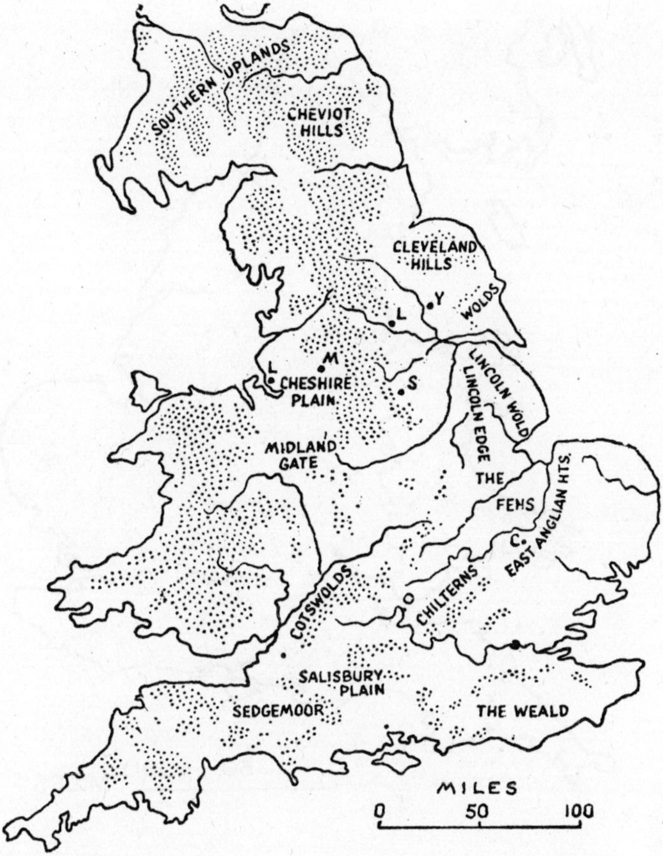

Fig. 1 (A). The Major Physical Regions of Southern Britain.

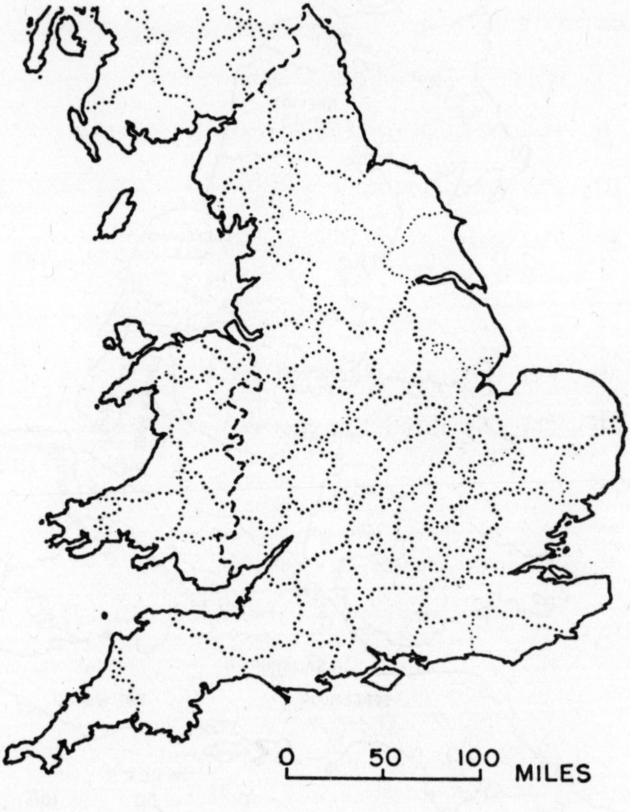

Fig. 1 (B). *The Counties of England and Wales*

CONTENTS

MAPS AND DIAGRAMS

CHAPTER I

WHAT IS HISTORICAL GEOGRAPHY?

I

THE NEED FOR DEFINITION

If this volume is to make any attempt to teach *Historical Geography* it is essential to try to make clear at the beginning what it is that we wish to teach and learn. This need is real and great for two reasons: first there is much popular lack of precision in the definition of the field of geography, and secondly, the historical geographer in particular must have the geographer's purpose and design always in mind if he is to use historical material successfully.

Most people would be hard put to it to say exactly what geography is. Botany is the study of plants; ichthyology the study of fish; mineralogy the study of minerals; physics is a little more difficult, it shades off into mathematics. But on the whole people know, or at least feel that they could find out, what the natural scientists study. More would hesitate to define history and many would dispute any definition given, but history is an old and eminently respectable subject and everyone knows what it is about. But geography . . . ? Some look upon the geographer as a kind of intellectual rag-and-bone man content to cull ill-assorted bits and pieces of information from many other disciplines. The bookshelves and the book lists of the geographer might perhaps seem to support the idea that he is a dilettante dabbling in many things, mastering none. Historical geography is a still greater mystery; few go further than a belief that it is about " old " maps, and perhaps concerns itself too with the tales of ancient mariners, medieval travellers and merchant adventurers. Some feel that it is an unsound attempt by geographers to explain history, and think that the historical geographer is most certainly trespassing and probably should be prosecuted. This is not so, the historical geographer is a geographer first, last and all the time, and therefore his first task is to teach himself geography. This is absolutely essential and the reader is referred to the other volumes in this series for study. In case, however, he has not these volumes all at hand or the time to read them all, an attempt is made in the first half of this chapter to set out very briefly the main principles guiding geographical study so that this book can be used intelligently without reference to its companion volumes, though the reader will thereby miss much.

II

THE STUDY OF GEOGRAPHY

What *does* the geographer study?

> " Biography is about chaps
> Geography is about maps."

Thus E. C. Bentley gives the geographer's theme: the geographer is concerned primarily with place and not with people. However, if the place is inhabited—and those places that interest us most are inhabited—then people are extemely important to the geographer; " Chaps " do much to make and alter maps. Historians, economists and anthropologists, those who think first of men, often use the word " geography " or " geographical " to refer only to the facts of physical geography. They talk of " geographical " facts meaning rivers and lakes, mountains and plains, or sometimes the less tangible conceptions of distance, area or space; and they talk of these in contrast with " economic " facts meaning perhaps industries or railways, or with " historical " facts referring maybe to frontiers or capital cities. The man in the street, however, does not think of the " place ", London, as a river terrace in the Thames valley, enjoying (or enduring) a western oceanic temperate climate. Neither does the geographer; to him too London, the place, is a town: the people, the houses and streets, the shops and factories, London river and the Docks, the Houses of Parliament and Buckingham Palace all come to mind, and come to mind as readily and perhaps more readily, than the latitude, the gravels and the weather. As one complex whole the geographer sees London and as one complex whole he must study it, and every other place.

The object of geographical study—the place: the area, the region, the whole earth's surface—is then no mystery, no vague concept, but a complex as real and as well defined as the plants of the botanist, the peoples of the anthropologist, the civilisations of the historian.

How then should geography be studied? Let one thing be said at the outset and borne in mind throughout, it must be studied at first hand, on the ground itself. What can be learnt of lands unvisited by the study of work done there is only appreciated in terms of comparison and contrast with lands known by personal observation and experience. "What should they know of England who only England know?"—precious little perhaps, but it is by that little that they learn the more wherever they travel and whatever they read. Geography like charity should begin at home. However, one can live in a place and be geographically almost wholly ignorant of it. The features of the landscape may be thoroughly familiar, yet have little meaning. Methodical study will bring a wealth of new interest; but how does one proceed? Since the work of the biographer is more familiar than the work of the geographer, a comparison of their methods, if the analogy is not pushed too far, may help. Neither the biographer nor the geographer is content merely to collect the facts about this subject nor to describe the outer form, both must strive to understand the facts they find and the form they see. The biographer must first examine the man himself, his body, his mind, his work. He is concerned with the action and re-action of body on mind and both on work. But he cannot neglect the factors external to the man, his relation to his fellow men as individuals and to society as a whole must be studied; these help to shape his body, mould his spirit, direct his work. Sometimes the biographer may go further and try to see his man, in spite of his personal idiosyncrasies and individual traits, as an example, characteristic or exceptional, of a more general type—

an example of his profession, his age or his country. Finally, and not always successfully, he may try to weigh the part that his man has played in his sphere; it is clear that there must be this aspect, each man by his very existence exerts an influence on all around him but the influence is hard to define and still harder to assess. A good biographer needs width of knowledge, he must be something of a physiologist, psychologist, economist, historian and perhaps philosopher. But he studies these disciplines, not as a specialist would, using the individual to check and increase knowledge of general principles, but to gain knowledge of the general principles as a help towards the understanding of the individual in whom he is interested.

The geographer groups the elements he studies in a very similar way: he studies the facts that in sum make the physical environment—the body of the place; he studies the living inhabitants—the plants, the animals and the men—the active agents, the spirits of his place; he studies the work done by these inhabitants in his place. He studies these elements not in isolation but as they act and react one on the other, one group on the other. He must examine his place in its world setting; he must consider not only the internal facts of the area but the facts external to it that none the less relate to it. He may then try to see the place, for all that it is like a man unique in character, as an example of a more general type. He may finally be tempted like the biographer to evaluate the influence of his place in particular terms in a particular case, or, a still harder, more hazardous and questionable task, in general terms in the general case. One thing is certain; the geographer cannot attempt this final task, if it is his, until he has made a full study and reached a thorough understanding of the place itself.

The geographer has a more difficult task than the biographer. The biographer is studying one man, a discrete unit which though related to other units is separable and distinct from them, and is of manageable size. The span of life of a man is short; the biographer is concerned with a manageable period. The geographer is concerned with the whole surface of the earth, an area not only of vast size but one and indivisible. He is concerned not merely with the surface of the earth in a limited sense, but with the atmosphere above the surface and the rocks and waters below the surface in so far as they help to explain the surface features and are a part of the environment of the earth's inhabitants. Further, the geographer is concerned with a vast period; some present features of the earth's surface date back not merely to early historical but to early geological times. The geographer therefore if he is to make any progress in his immense task must concentrate his forces to pose the most vital questions, to marshal the most rewarding facts to answer them.

The Essential Geographical Problems

The two essential questions that the geographer must ask are " where ? " in order to find out how things are arranged in his place, and then " why there ? " in order to appreciate the arrangement that he has found.

The question " where ? " raises the problem of distribution and it is not always as easy of solution as it would seem. It is astonishing how difficult it is sometimes to find out exactly where things are. Some things have no clear-cut permanent position. a warm damp air mass, a herd of grazing elephants or a group of pioneer farmers for example. Sometimes it is difficult to make sure that " things " are really comparable. When is a desert not a desert, a working coal mine not working, an Aryan not an Aryan ? These are pertinent questions. Then the " things " must all be there at the same time ; it is not always easy to find out if this is so or to decide how great a variation in time can exist and yet the things be considered contemporary.

The study of distribution is of first importance to the geographer because it reveals the variation there is between place and place. Most things are not scattered over an area at random but show definite arrangement in space ; their distribution shows that what may be called " geographical pattern " exists. The geographical patterns of many phenomena compared one with another serve to bring out the characteristic features of the place ; serve to establish its geographical unity and individuality. But the geographer is not content merely to recognise this individuality, he must strive to understand it and this brings him to his second problem, the problem of location.

The problem of location, the search for the answer to the question "why there?" is often very difficult. A fully satisfactory answer is but rarely found, but the search is fascinating and, if honestly and carefully pursued, rewarding. The first step is to compare the various geographical patterns that emerge as the result of the study of the distribution of geographical facts. The patterns are nearly always complex like those of a fine Persian rug; what is more, there are usually several alternative patterns to be seen. Look once and the mountains and the plains seem the dominant masses; look away and look again and the cultural and political affinities of the peoples seem to give the leading lines, and may offer another design. The patterns of some phenomena seem to vary independently of one another; for example, there would probably be no similarity in the pattern of let us say the distribution of coal-mines and the frequency of summer thunderstorms. Other patterns vary either directly or inversely one with the other, and much can be learnt from studying them. For example, a correlation in pattern between a low standard of living, a low density of population and a relatively fertile land, as for example in some of the coastal lowlands of Macedonia, might suggest that for some reason man was not very efficient in the area. A further correspondence in pattern with, say, areas of stagnant water, a high density of that mosquito which most commonly carries malaria, and a high incidence of sickness might suggest one reason at least for the first correspondence noted. It is always necessary to remember that things may not be as they seem. In Britain a correlation in pattern can be seen on distribution maps showing areas of high rainfall, high relief, a high percentage of sheep farms, a high percentage of Celtic speakers and a high incidence of dark hair and dark eyes among the people. But to proceed to argue that the Celts were a dark folk pursuing a pastoral economy who preferred to settle in wet mountain districts is to construct a whole series of fallacies. There are of course links between these facts, and before all of them are forged a long and complex chain is formed that incidentally leads back to prehistoric time, to the very earliest phases of the colonisation and settlement of our island. As the search for cause and effect proceeds the realisation of the complexity of the interaction between geographical phenomena grows.

It would be exceedingly dangerous to believe that all the links in the chain of cause and effect can be found by a study of distribution patterns alone; distribution patterns only suggest some of the relationships for which explanations must be sought. The factors of chance, material and immaterial, have by definition no pattern but often play an important and sometimes an all-important part in the problem. However, there is a limit to what chance can do; a tomato seed dropped by chance in the Arctic will not germinate, though by design, by providing soil, heat and light artificially in a greenhouse, it would grow.

It is sometimes said that modern scientific techniques have made nonsense of the study of location. This is not so. Modern scientific techniques have vastly increased man's power to adapt himself and his work to his environment and to adapt his environment to his needs. He has far greater freedom of choice among the opportunities that his environment offers to him and this has made the study of the problem of location more difficult because more complex. For example, the Russian experiment in the development of arctic lands has made it no longer sufficient to study the relationship between the biological needs of the plants and the climate of the area in examining the distribution of the crops of Siberia. The breeding of special plant strains, the provision of an artificial climate and soil, the costs of production under these conditions, the market demand, the possibilities of and the cost of transport, even political ideologies may all need investigation before the pattern of Soviet agriculture in the Far North is understood and its implications even faintly appreciated.

The Fundamental Geographical Facts

To find the essential features of a place and to understand their synthesis it is necessary not only to ask the right questions "where?" and "why there?" but to ask them about the right things. Geographical facts are innumerable but they are not all equally important.

A geographical fact is any fact that can be mapped on the earth's surface, any phenomenon that has an address. To have any meaning however in the study of place a geographical fact must differ in kind or number from place to place so that it serves to distinguish one place from another. To be important a geographical fact must be one that not only concerns something that varies from place to place so as to give a well-defined geographical pattern, but one that is itself a conspicuous feature in the landscape or exerts an influence, either as a passive factor or an active agent, upon other conspicuous features. To quote Sellar and Yeatman: "Every right-thinking person will recognise at once that such things as Sponges, Tundras, Tablelands and Cameroons are exquisitely geographical, whereas Sponge-cakes, Tumblers, Table-napkins and Macaroons are not geographical at all."

The most fundamental geographical facts, or rather groups of facts, are those relating to world position, climate, land and water forms, soils, plants, animals, men and the economic, political and social activities of men in organised communities. These groups of facts are each for their own sake the primary subject of study of some specialist and the geographer must learn about them from the specialists concerned. Sometimes he may have to teach himself as the specialist may not be interested in the particular things the geographer needs to know; but the geographer even in this case must be very clear that he is primarily a learner not an investigator in these fields. He must learn of the chemistry and physics of the atmosphere and the behaviour of air masses from the meteorologist in order to understand the elements of climate and the way in which they vary over the earth's surface. He must learn from the geologist about the properties of rocks, the earth forces and the resultant earth structures in order to understand the way in which the rocks react to the agents that shape the landforms. He must learn from the botanist, zoologist, physiologist, anatomist and anthropologist about the structures and way of life of plants and animals and men if he is to understand their geographical distribution and the effects they have on the landscape. He must learn of economic principles, political theories and historical events from the economist, politician and historian if he is to begin to understand the way in which societies vary from place to place and the complexity of their direct and indirect effects on their natural environment. Though the geographer is concerned primarily with material facts he must realise the importance to him of the world of intellectual ideas. Philosophical concepts, moral judgments and religious beliefs often find strong geographical expression and the geographer must be prepared to seek help with what are perhaps his most profound problems from the psychologist, the philosopher and the theologian.

The geographer, having learnt about the facts he uses (in practice of course he learns as he uses), then applies his knowledge to his own investigation. His investigation may be at first a work of analysis, a teasing out of the relationships between the different sets of geographical facts, but in its end it is a work of synthesis, a weaving in of all the strands to give as nearly as possible a complete appreciation of the whole.

His work is to examine the inter-relationships of all the elements of all the major groups of geographical facts, not only as they exist between element and element, but between element and group, and between group and group.

Most of these inter-relationships in whole or in part concern other students also. There are no frontiers, no iron curtains, no closed shops in knowledge, in so far as one student can be distinguished from another, it is his end, his aim and purpose, that distinguishes him. The botanist, for example, studying a rice field examines the whole complex of the inter-relation of climate, soil, relief, and probably cultivation methods, and therefore labour and market conditions too, but he examines them to throw light upon the growth and development of the rice plant as a contribution to his knowledge of plants in general. The anthropologist, sociologist and historian may be interested in the whole gamut of inter-relations that concern the geographer, though they as a rule are content to take the complex of the natural environment as a whole and do not concern themselves with the detailed synthesis of the different elements within it. But it is not this but their purpose that distinguishes them. Their studies are undertaken to throw light upon the character of society and civilisation; their study of the relationship of man to his environment is only one step towards this end, one part of their immense range of interests. The geographer in contrast is concerned with all the inter-relationships tangible and intangible that directly or indirectly have material expression in the place. He is concerned not essentially with individual features as such but with the complete assemblage. His whole study, the complexity of which has been here but faintly suggested, is undertaken for the purpose of elucidating first the intrinsic qualities of each place, each region, and secondly the links that bind inseparably the unit parts into a single place—the whole world.

No one of us can aspire to be the complete and perfect geographer equally expert over the whole range of the discipline as applied to all parts of the world. All of us must specialise in the study of a particular region or a particular topic, few of us can hope to contribute new knowledge except within a very limited range as applied to a very small area. However, each of us must see his work, if it is to be fertile, as a part that fits neatly into the general theme, however minute a part it may be of the vast whole. All of us then must, while we cultivate our own tiny plot, take time to contemplate our neighbours' plots and learn from them so as to obtain gradually a better and better general knowledge of the geography of the whole surface of the earth.

III

WHAT IS HISTORICAL GEOGRAPHY?

What then is Historical Geography ? It may be useful to clear the ground by stating first what it is *not*.

Historical geography is sometimes used as a title for the study of the development of geography as a discipline. The study of geography is traced from the time of the Greeks, when geography included all the disciplines that would now be regarded as belonging to the Natural Sciences, to the present time when it is once more a scientific study distinct from, but in close contact with, the daughters it has launched. In the Early Middle Ages scientific interest in geographical facts became entangled with philosophical and religious concepts of the universe and the real and the metaphorical became hopelessly confused. The voyages and explorations of the Great Age of Discovery revived interest in the real world but for long new facts came in so rapidly that there was little time to do more than record them; scientific study was given only to the examination of separate groups of phenomena, not to their aggregate effects in the world as a whole. There follows a barren period when geography seems to be concerned only with matters proper to the gazetteer, and unadorned and unmeaning lists of capes and bays, towns and their manufactures were presented to the hungry student. The nineteenth century brought revival; Ritter and von Humboldt laid the foundations of the "new" geography (with links in attitude at least to the "old" geography of the classical world) and their work was followed up, for it shared the general interest in environment aroused by Darwin. In our own day our study owes much to the great names of De Martonne and Vidal de la Blache in France, Hettner in Germany, Herbertson and Mackinder in Britain, W. M. Davis and I. Bowman in America. The development of geographical ideas is an interesting study and those who wish to follow it will find a good introduction in R. E. Dickinson and O. J. R. Howarth, *The Making of Geography* and a full discussion in R. Hartshorne, *The Nature of Geography*. This study is not properly included as part of historical geography: it is better termed the History of Geography, it has its parallels in the History of Medicine, the History of Botany, the History of History come to that.

Many books with the title *Historical Geography* would be better entitled *Geographical History* for they are concerned essentially not with the place but with the civilisation. Much that is written on this subject is bad, broad generalisation on an insufficient basis but H. B. George, *The Relations of History and Geography*, though old-fashioned now, is suggestive, and J. M. Thompson, *The Historical Geography of Europe* is a very useful outline. It would seem that the attempt to examine historical events in relation to their geographical setting is an exercise best left to the historian. It is for the historian to urge the historian to look at the ground and the map; though the geographer may perhaps help to construct the map, and aid in interpreting it.

Historical Geography : the Geography of the Past

Historical Geography is, simply stated, a geographical study of any period in the past for which a more or less ordered and dated sequence is established in human affairs. There is no need to distinguish sharply between history and pre-history. Emphasis is placed on the *geographical* study. The facts studied and the methods of study are those briefly discussed in the first part of this chapter. It is often assumed that the broad distinction between the geographer and the historian is that the historian is concerned with the past, the geographer with the present, or that the historian is concerned with " periods ", the geographer with " countries ". These are false assumptions ; the historian is not concerned only with the distant past, he can and does apply his method of study to the immediate past. No one disputes the existence of contemporary history, the historian does not become a geographer when he studies the present. The historian is concerned with the world as a whole, he can for convenience of study divide his world into regions ; he can write a history of Peru or India or France and no one considers that he thereby becomes a geographer. Similarly the geographer can divide his world into periods ; he can study the geography of the nineteenth or the ninth century and remain a geographer. There is much in common between the historian and the geographer, both are attempting to see the pattern in a multitude of facts in an effort to appreciate the world about them, but there is a fundamental difference in outlook between them. The " world " to the historian means civilisation ; the " world " to the geographer means the surface of the earth.

The historical geographer, if his attitude of mind and his methods of study are the same as those of all other geographers, is at a much earlier stage of his study than many of his colleagues are. The field he has to till is almost virgin. The answer to the question " what was this place like then ? " is still in most cases " we do not know ". Therefore in studies of historical geography, the first stage of study, the establishment of the geographical facts, receives more attention than in many other geographical works. It is necessary to determine the facts of distribution before discussing the problems of location. However, the art of synthesis must be constantly practised, not only in order to understand the patterns appearing but also sometimes to help to find them. Often the only bits of information that come to hand are like isolated pieces of a jigsaw and, by the use of his knowledge of the way in which geographical facts interrelate, the historical geographer may find a clue that will lead him to other pieces and guide him in their arrangement. The method must of course be used with integrity, as a torch to help search and interpretation, not as a substitute for them.

The historical geographer then must be a good physical and biological geographer. Physical and biological geography form the basis of all the rest of the study, for if the relationships between man-made features and the natural environment are to be assessed, the natural environment must of course be fully understood. There is no need in this book to discuss the interrelationships that determine the geography of the physical environment. The fundamental relations between position, climate and relief are unchanging in time. They are not peculiar to the field of historical geography and therefore it will be assumed that the reader has gained, or will gain, this knowledge elsewhere. The second volume of this series R. F. Peel, *Physical Geography* provides an introduction and P. Lake (edited J. A. Steers), *Physical Geography* is a new edition of an invaluable classic. Both contain useful bibliographies to guide further reading. Similarly the fundamental principles governing the relation between living things and the physical environment are unchanging in time; the facts change but the character of the links between them and the processes that alter them remain the same. For a discussion of the geography of the biological environment, especially emphasising the geography of man, the reader is referred to the third volume of this series M. S. Anderson, *Geography of Living Things*.

The forces of nature are the large-scale agents of geographical change; they work slowly, but inexorably, over the geological time-scale to bring about veritable geographical revolutions: mountains become plains, lands become seas, ocean deeps are replaced by island arcs and mountain chains. Man in comparison is a small-scale agent of geographical change but he works rapidly to bring about irreversible changes that modify, but modify rather than revolutionise, the geographical scene. All geographers are much interested in these modifications, perhaps because " the proper study of mankind is man ", perhaps because of vain pride in " the vulgar works of man". The historical geographer is almost entirely preoccupied with them: if a study of the place in the past is to be valuable it must differ from a study of that place in the present, and for the most part it is the work of man that brings about significant change in historic time.

The Purpose of Historical Geography

The historical geographer is thus concerned mainly with the social, economic and political geography of an area at some past time. The study is undertaken for two reasons: it has an intrinsic interest of its own, and it provides the necessary background for the geographical study of the modern world. The historical geographer *qua* historical geographer is not concerned with the *survival* of geographical patterns or the *evolution* of the geographical patterns in time, but with the establishment and study of their design at any *one* particular time. He emphasises especially those features of place most characteristic of that particular time. The value of his work *qua* geographer however lies, on the one hand, in the fact that some elements of geographical design that develop in response to passing conditions are extremely stable in their form or long lasting in their effects, and the understanding of the present demands the study of the geography of the period of their establishment and development. On the other hand there is also value in the fact that other elements are rapidly changing and the study of their change and evolution may throw light on general principles that determine their geographical pattern, principles that the human geographer no less than the physical geographer is ultimately striving to establish.

The Scope of Historical Geography

The great themes of the historical geographer concern then the long-lasting, stable elements of the geographical scene: the distribution of peoples, the patterns of settlement, the soil and vegetation zones that develop as fields and pastures replace forest, marsh and moor—the study of these lead back in most areas to much earlier ages than the present. The great themes are also those rapid changes in a region that man's increasing ability to adapt himself to his environment and to modify it bring about: the changing patterns of agriculture, of industry, of communications and trade, of population densities, and the resultant changes in the evaluation of natural resources and of space relations. The historical geographer is concerned too with the development of political and administrative units in so far as their changing areas and frontiers and their changing policies have geographical effects. Further, he cannot consider the internal features of a particular place at a particular time without considering the external world of which it is a part, but he is not concerned with the world as it is today but with the then known world. To study medieval Europe in isolation from India and China would be absurd, but to assume that the strong sea links between these areas were already forged then would be equally absurd. Thus the study of the progress of geographical discovery and the growth of geographical knowledge must always be an essential part of the historical geographer's field.

IV

THE PLAN OF THIS BOOK

It might be expected that a series of regional studies, each one devoted to a different period—Roman Britain, Renaissance Europe, Colonial America, Africa in the Nineteenth Century, for example—would form the different chapters of this book It may be that this is how the book should have been written, but it was felt that even had the author the knowledge to attempt it, the reader would not gain much help in pursuing his own study from widely selected but isolated examples, however carefully chosen. Treatment would be too brief in the space available. It is judged that the beginner is most likely firstly to lack appreciation of the geographer's need to consider the place in the past at all and secondly to be unfamiliar with the sources of information that can be used to establish, however baldly, something of the geographical patterns and space relations of a past period. Each chapter is therefore concerned, not with a particular area at a particular time, but with a particular theme, and an effort is made in each to show why the topic repays or needs historical geographical study; to indicate the sources available for its study; and to give some examples to show how to handle these sources. This method is adopted in spite of the danger that the avowed purpose of the historical geographer to re-create as fully as he can the geography of an area in the past will lose emphasis. If the reader investigates each topic for himself along the lines suggested for his own locality and at a period of interest to him, he will thereby assemble the essential material from which a regional synthesis can be essayed. The technique that the geographer practises in attempting a synthesis of the separate elements to create an understanding of the living reality of a place is, as this chapter has been written to show, the same whatever the region, whatever the period studied. The material to be studied and the sources of that material, not the method of regional synthesis, is what is peculiar to historical geography and therefore it is upon material and sources that emphasis falls in this book.

The treatment of these topics in the following chapters may often seem local, even parochial, in scope: this is not a contradiction of the precept that geography is concerned with the earth's surface as a whole. It is a golden rule to teach by example and it is an even better rule to learn by practice. If these rules are to be followed here, the material to study must be available. The records and the books must be to hand, and, even more important since the time may be distant, the place must be accessible. World-wide travel is a dream for most of us, we must be content with the home ground. But, if the art of investigating the local facts of historical geography is practised, it becomes easier to appreciate how studies are made elsewhere by others and how they are pieced together to fill in and check the broad outlines of our knowledge of the world. To stare at the village pump in order to look more intelligently at other pumps is not a waste of time. Finally, in attempting to teach historical geography, the teacher is faced all the time with one overwhelming difficulty: his own colossal ignorance. He can but use the material that he has to hand as example, and hope that example will suggest practice on the material that lies to the hand of the learner: teacher and learner must "adventure" together on their voyages of discovery.

CHAPTER II

THE DATA OF THE HISTORICAL GEOGRAPHER

IN the first chapter it has been emphasised that the historical geographer pursues his study of the place in the past in exactly the same way as he would as a geographer study the place as it is today. The would-be historical geographer has in fact been warned that first and foremost he must become a good all-round geographer. It may be as well to begin this chapter by emphasising that the greater part of the data that he uses and the facts of place in which he is interested can only be appreciated and understood if their period is appreciated and understood too. The historical geographer must then be also a continuous and diligent *pupil* of the historian.

To acquire as wide a historical knowledge as possible should then be every reader's aim ; this is an underlying assumption throughout the book, but no discussion of sources can begin without stressing it. This is not the place and the author is not the person to give advice on how to " Teach Yourself History". A. L. Rowse in his book *The Use of History* introduces this subject and gives a useful bibliography to guide further steps.

The historical geographer desires the facts about his place in the past to fulfil three conditions : he wants them to be strictly localised, to have a fixed position in space ; he wants them to be well, that is evenly, distributed over the area ; and he wants them to be representative of their kind since he seeks examples of the rule rather than the exception within the area. His demands are exacting and not easy to fulfil. It must be realised that the facts revealed by other specialists may not, in fact almost certainly will not, fulfil all the conditions that the geographer requires. There may be nothing to show exactly *where* the manor-house or the mill stood, exactly *where* the arable lands, the pastures and the woodlands lay, although there may be records of their owners and their values over long periods. Information may be abundant about one part of an area studied and absent for another part. The exceptional and the disastrous may be recorded while ordinary everyday practice may not. The imperfections of the data available to the historical geographer are in part inherent in the material, and in part in the specialist study of the material. Much that the historical geographer would like to know of past conditions has left no overt traces and no records to find. Much of the data that exists has not yet been examined in the way in which it will yield most geographical information: the facts the archaeologist and historian like best and therefore seek first are not always those the geographer most wants to know. This latter imperfection will improve with time as more knowledge is accumulated and as there is more co-operation between geographer and historian. The geographer can stimulate the historian to search for the facts he needs by asking the questions he most wants answered. The historian shows signs of greater interest in the geographical facts and of greater awareness of his own need of them to furnish his stage.

What material can be used to answer the questions of the historical geographer depends very much upon the period of the study. Even for the period of the first colonisation and settlement in "new" countries there are documents in plenty for consideration. The facts relevant to the early phases of the geography of the "old" world will not for the most part be found in written records at all. The myths and legends embodied in early written records tell most often of war and battles, of kings and counsellors, but behind the stirring events the ordinary everyday doings of the mass of common people, tilling the land, herding animals, cutting forests and draining marshes, building villages and towns was going on unwritten and unsung. Conquerors and kings undoubtedly prepare the way for geographical change but peasant and artisan for the most part bring it about. How can the geographical work of these humble folk in the distant past be traced? All kinds of indirect evidence, a bit here and a bit there, must be pieced together to give clues. There are many gaps and many guesses and the searcher must always work with an open mind. He must cast his net widely to find his data; he must search the ground itself as it now is for traces of the features of the past geographies, and he must consult those specialists concerned with digging up the past literally and metaphorically, in particular the historians of material finds, the archaeologists and the architects, the historians of language, the etymologist, as well as the historians in the usual sense, the students of the written records.

The beginner should search out the sources available in his own area and try his hand at collecting evidence around his own doorstep ; his attempt should be made not primarily in the hope of reaching satisfactory conclusions or making new discoveries, but in order to see what kind of material is available and what kind of deductions can be made when and if enough is known ; though the veriest beginner may always be lucky in finding through his local knowledge some new fact to add to the common store.

II

The most important sources of material for historical geographical study are :—archaeological finds, i.e. the material remains of earlier societies ranging from skeletons and small ornaments, tools and weapons, to houses, forts, castles, temples and churches, including buildings more usually thought of as architectural monuments than as archaeological remains ; place names ; statistical and quasi-statistical material, perhaps unreliable and certainly incomplete in earlier periods, becoming almost overwhelming in abundance in the nineteenth century ; acts of Parliament and the reports of Parliamentary commissions ; contemporary histories and geographies or topographies as they were more often called ; contemporary journals and travel books ; contemporary road books and maps.

Archaeological Evidence

The study of archaeological finds and of the architecture of our buildings does much to supplement the written record, especially in the investigation of the geography of early periods. The identification and dating of archaeological finds and architectural remains is often very difficult and is always a matter for the expert. Dr. Graham Clark gives a survey of the aims and methods of the archaeologist for the general reader in his book *Archaeology and Society* and the historical geographer wishing to use archaeological material should read it. Books on architecture are legion but two picture books published by the Architectural Press, F. Chatterton, *English Architecture at a Glance* and J. M. Richards, *A Miniature History of the English House*, will provide a useful introduction for those with little knowledge of architecture. Two rather larger books are indispensable to the geographer: T. D. Atkinson, *Local Style in English Architecture*, and A. Fellows, *The Wayfarer's Companion*, which treats of both archaeology and architecture.

Where will the geographer find the information the archaeologist and architect has to give him? He may make a preliminary search on the Ordnance Survey maps; even on the Popular Edition 1 inch scale the known sites of importance are marked, and they are shown in more detail on 2½ in. and 6 in. O.S. sheets. He may next turn to air photographs which show up, by their extremely sensitive emphasis of light and shade, very slight variations of slope, water content and therefore colour of soil and crop which often cannot be appreciated at ground level. The outline of former earthworks and buildings, the run of disused trackways and boundaries can often be traced by these variations, though their identification needs training and practice. The Ordnance Survey is now publishing a series of air photographs on a scale of six inches to the mile, four photographs covering each 2½ in. sheet. The publications on local history should next be searched for articles about the features seen on the ground or on the maps; and discussions and excursions led by local experts are if available most helpful.

In studying the distribution patterns that archaeological finds and architectural remains reveal and in making deductions from them the geographer must remember the limitations of the evidence. Where the remains of man of a given date have been found *in situ* undisturbed, certainly man was present there then, but the converse is not necessarily true. Where no remains are found there may well have been settlement. It is necessary to remember that only that which by chance has survived can be found; all the potsherds and bones ground to powder, all the wooden tools and weapons decayed or burnt, all the sites built over, all the fields ploughed over leave little or no trace. Further, the number of finds and their distribution may reflect not only the density of the settlement of their period, not only the survival rate of the objects, but also the industry of the archaeologists in the neighbourhood. Other areas not yet explored may prove to be just as rich in finds. The apparent liking of early man for Oxford river terraces and Cambridge gravels reflects to some extent the liking of Oxford and Cambridge students for excavation.

Again the date, the size and the ornamentation of buildings will frequently provide clues to the distribution of wealth and thus indirectly to the distribution pattern of population density at different periods. This evidence has limitations too. The existence of a church with a fine Norman tower, for example, pre-supposes a group within the area in the late eleventh or twelfth century sufficiently numerous to provide the wealth and labour to put up the building. Churches of whatever period, and to some extent houses too, reflect by their size the wealth of the group but not necessarily their number; a church or house is not built as large as is necessary but as large and as magnificent as the people can manage. This fact needs to be remembered by the historical geographer. Further, of course the absence of buildings of an early period does not imply an absence of settlement at that period. In many places as wealth accrued an early building considered small, unfashionable and inconvenient might be pulled down and a more imposing one built in its stead. The magnificent perpendicular churches of many of the villages of southern Suffolk, Clare, Long Melford, Lavenham, Hadleigh, Nayland, Dedham and the rest point not to settlement and colonisation of the area for the first time in their period but to the wealth of the district then, resulting in large part from the flourishing woollen industry that grew up here in the fourteenth and fifteenth centuries.

Place Names

Place names carefully used provide much valuable evidence. Names of places survive, even though much altered, from very early times; they are precisely located, numerous and well distributed and thus fulfil the very exacting requirements the geographer demands of his data. A study of the language of the place names will give some clue to the people who performed the christening ceremony. The distribution of the languages of the place names gives evidence of the movements of peoples, the areas of their colonisation, and their relation to earlier and subsequent groups in those areas. The date of the place names tells something of the date of the settlement, or at least of the date of the establishment of the name as it has survived. The meaning of the name may tell of the people or their social organisation or of the nature of the environment; all facts relevant to geographical study.

For information on the language, date and meaning of place names the geographer must ask the etymologist and philologist. E. Ekwall, *The Oxford Dictionary of English Place Names*, E. Ekwall, *English River Names* and A. Mawer, *The Chief Elements Used in English Place Names*, are indispensable dictionaries and word lists. The introduction to the *Dictionary of English Place Names*, and Vol. I Part I of the Place Name Society Publications entitled *An Introduction to the Survey of English Place Names* edited by A. Mawer and F. M. Stenton are lucid discussions of the nature and scope of place name study. The geographer must needs make constant reference to these books; for detailed local study he will wish to consult as well the county volumes published by the English Place Name Society; many volumes are available but the whole series is not yet complete. Beginners reading place name literature are apt to be overwhelmed by the wealth of etymological and philological detail, but if they persevere they will soon become familiar with a sufficient vocabulary of elements common in place names to enable them to derive much profit and more amusement from a study of signposts. In Britain the elements of most place names belong to one of three languages: the language of the pre-Roman inhabitants sometimes called Celtic but more properly called British, the language of the Anglo-Saxons usually called English, the language of the Vikings, Scandinavian. One soon learns to recognise elements that are characteristic of each: for example,

Pen, a hill as in Pennard, *Tre*, a hamlet as in Tremaine, *Avon*, a river as in Avening, are very common British elements; *—ing* (ingas), a people as in Hastings (the people of Haesta), *Reada*, Reading (the people of Reada), *—ham*, a village as in Tottenham, Chatham, *—ton*, a clearing round village hence village as in Bolton, Edmonton, are very common English elements; *—by*, a village, town as in Whitby, *—thorpe*, a hamlet as in Felthorpe, *toft*, a homestead as in Lowestoft are easily recognised Scandinavian forms. Many names are descriptive of the place: ox-ford, wood-bridge, black-pool, Peter's-field come immediately to mind. Many names tell something of the natural environment: those ending in *fen*, *-moor*, *-wood*, *-grove* contain elements of obvious meaning; *-field* and *-leah* (ley) are very common elements and seem to mean clearings or open spaces in the forest or wood. Oakley, Ashley, Haseley tell of the kind of wood, and Staveley, Yardley, Stockleigh tell of the products of the wood and hint at the economic geography of the time. A word of warning is necessary: however obvious the meaning of a place name may seem it is *never* safe to guess it; it is always necessary to look it up. Even the most innocent-looking words can be traps. Oakford, Aldreth, Haslemere, Elmstead all have first elements which are what they seem—names of common trees—but Elmstree is the tree of an old English gentleman named Æthelmund. Woodborough, Woodbridge, Woodhouse are all innocent and give a feeling of safety to the amateur who conscientiously looks up each in turn, but Woodsford is there to catch the unwary, it means Wigheard's ford and has nothing to do with 'wood'. Westley Waterless, a village near Cambridge on the chalk, would seem to be appropriately named, but when earlier forms of the name are seen the meaning is almost the opposite of what it seems; Waterless is Waterlees, i.e. water meadows, and the *-ley* has also been tacked on to the West and is of course redundant in meaning. The normal vowel changes that are characteristic of the evolution of the language, mispronunciations sometimes consequent upon the attempts of 'incomers' to get their tongue round a foreign name, and popular, and therefore often false etymology, have changed many names in the course of centuries almost beyond recognition. The changes themselves are often of value and suggest clues to the historical geographer. The Venerable Bede gives us the earliest recorded form of York, he spells it Eburacum. This seems to be a Latinised form of *ebur* which, according to some authorities, is a Gaulish word meaning 'yew', according to others, a British personal name. To the Anglo-Saxon invaders Ebur meant nothing, but they had a word for boar which sounded rather like it—*eofor*, so by popular etymology or by mishearing of the unfamiliar as the familiar, Ebur became Eofor. A suitable ending for a small settlement was *-wic* which also meant pasture so Eburacum becomes to the Anglo-Saxons Eoforwic, a sensible name—the pasture of the boar and by association the hamlet at the pasture of the boar. Then came the Scandinavians; Eofor was rather a tongue twister to them and they simplified it to Ior. *Wic* meant nothing to them but *vìk* meant a harbour, a haven, and since the place was at or about the head of navigation of the Ouse it became Iorvìk and hence York. Here epitomised in the name is the transformation of the humble dwelling of the unknown British resident Ebur into a hamlet in a clearing, into a port on a busy river, and the arrival of the three major cultural elements in the English nation, British, English and Scandinavian.

Written Records

However useful the evidence of archaeological material and place names, the great bulk of the data used by the historical geographer is found in the written records of the past. Statistical material that will give reliable and full quantitative data is of course not to be found for any country until the closing decades of the nineteenth century, but earlier national and local records sometimes yield statistical or quasi-statistical material which, fragmentary though it is, can be used to give a qualitative comparison between place and place.

The English historical geographer, for example, has one great record in the Domesday Survey that is without parallel in the information that it affords for so wide an area at so early a time. It owes its importance largely to two facts, its date and, for all its maddening variations and curious gaps, its uniformity and completeness. The survey was made in 1086, at the end of the period during which bands of settlers, some hard on each other's heels, some arriving after a long interval, crossed the Narrow Seas and established new homes in the British Isles. It was made also at the beginning of the longest period of constant development; some of the main lineaments of the geography of Norman England survived with very little modification for centuries and are therefore so deeply engraved that they show on our landscape even today. The geographical scene that the Domesday Survey helps students to establish serves then both as a sum of the earlier geographies and as the starting point of the later ones. The survey was carried out for all parts of England sufficiently well settled at the time to be brought within the seignurial organisation of the manorial system that linked each man directly to his superior lord and indirectly to the King. The lands beyond the Tees were hardly "English" in a political sense, they belonged to the great Kingdom of Northumbria that once lay across the Border from Tees to Firth of Forth. Similarly the settled parts of Cumberland and Westmorland lay within the Pictish kingdom of Strathclyde which, stretching north to the Firth of Clyde between Northumbria on the east and the wilds of Galloway on the west, was also outside the jurisdiction of the Norman Kingdom. The Domesday Book contains no records of these border areas. It covers, however, all the most densely settled and wealthy parts of Britain, the eastern, southern and western counties of England, the Midlands, and the three Ridings of Yorkshire; Lancashire as yet sparsely settled, was surveyed, the north as part of Yorkshire, the south as part of Cheshire.

Among the facts recorded are many that are relevant to the study of place; more would be perhaps if we understood the terminology of the surveyors better. According to the Chronicler, at the Gemot of Gloucester in 1085 "the king had muckle and deep speech with his wise men about this land, how it was set, and with what men. Then he sent men over all England into each shire and let them find out how many hundred hides were in that shire, or what the king had himself of land or cattle in those lands, or what rights he ought to have in the twelve months from that shire. Also he let them write how much land his archbishop had, and his bishops, and his abbots, and his earls and though I tell it longer, what or how much each man had that was landsitting in England in land or cattle and how much it was worth. So very narrowly did he let them speir it out that there was not a hide nor a yardland nor—it is shameful to tell though he thought it no shame to do—so much as an ox or a cow or a swine was left that was not set down in his writ: and all these writs were brought to him afterwards."

Government scribes at Winchester made copies of the returns in two great volumes, the Little Domesday covering the counties of Essex, Suffolk and Norfolk and the Great Domesday covering the rest of England. These volumes survive as one of our greatest national treasures, the Domesday Book.

The mass of data contained in the Domesday Book must be plotted on a map before it can be used effectively for geographical study. This involves an immense amount of labour, and also a nice judgment and much ingenuity if the method of plotting is to allow for variations in expression of the same items from county to county and even within one county, and is not to obscure gaps in the data or to evade difficulties created by imperfect understanding of the meaning and significance of some of the items recorded. Under the editorship of Professor Darby, the *Domesday Geography of England* is in course of publication.

A transcript of the Domesday Book for all counties was published by the Records Commission in the nineteenth century and a translation is included in the Victoria County Histories, but not all the county volumes are yet published. This translation where available is the most convenient form of the survey to use, and is in fact the only one beginners can use, for in it the specialist work of identifying the Domesday vills has been done and the names of the places appear in their modern form.

Even in a country where documents are as well preserved as in England, in the centuries between the Domesday Survey and the first census return of 1801 records that will allow the analysis of geographical distribution of people, their wealth and their occupations are few and fragmentary; one kind of record may exist for one year for a particular county or even parish but be missing for the same county in later years or for the neighbouring counties in the same year. Beginners can rarely use these records in their original form, but extracts or digests of some of them have been published in books or local journals and to appreciate the nature of the material readers may like to look at a few. For example a considerable number of custom and subsidy accounts for English ports survive. These accounts can be used to throw light on the geography of English trade at different periods. The original accounts are for most part in the Record Office, but some idea of the nature of the figures of the day-to-day account can be obtained from the extracts from them in N. S. B. Gras, *The Early English Customs System*, and an idea of the digested form of these as they were " enrolled " for preservation in the Exchequer Records may be gained from the tables in *Studies in English Trade in the Fifteenth Century* edited by E. Power and M. M. Postan. For information about similar statistics for a later period readers should consult G. N. Clark, *A Guide to English Commercial Statistics* 1696-1782. Taxation returns are sometimes useful in the study of early industrial geography. For example, records of a tax levied on cloth, sold in the public market, the ulnage accounts as they are called, survive for scattered years in the fourteenth and fifteenth centuries and allow some tentative conclusions on the location of the woollen industry in the medieval period to be made. H. L. Gray, " The Production and Exportation of English Woollens in the Fourteenth Century ", *English Historical Review*, Vol. XXXIX. 1924, shows the use of this material, but E. M. Carus-Wilson "The Ulnage Returns; a Criticism ", *Economic History Review*, Vol. II. 1929, points out that these returns must be used with caution. Records of a Poll Tax of 4d. per head of the adult population of England in 1377, of the Hearth Taxes collected in the late seventeenth century and of the Window Taxes levied in the eighteenth century survive for most counties and many towns, and if they are plotted give some idea of the distribution of population or wealth and therefore indirectly of the pattern of regional variation in population density at their date. Examples of maps made from some of these figures will be found in H. C. Darby (ed.), *The Historical Geography of England before the Railway Age*.

Manorial records—extents, rent rolls, bailiffs' accounts of crop and stock yields—often give statistical material of particular places for particular years but, except for a few big estates where records have been exceptionally well preserved and refer to many widely scattered farms, the material is too strictly localised to be very useful to the geographer.

In the nineteenth century statistical information becomes abundant and increasingly more accurate. The first census was taken in 1801 and from the middle of the century information is recorded about the age, the place of birth and the occupation of those enumerated. The Board of Agriculture founded in 1793, and the Board of Trade set up in 1696 issued first of all reports of Commissions at irregular intervals and, later, Year Books. There is for the nineteenth and twentieth centuries no lack of statistical information for geographers to use, in fact there is almost an *embarras de richesses*. A discussion of this later material will be found in the fourth volume of this series W. S. Thatcher, *Economic Geography*.

The records of Parliament sometimes contain much material which gives useful geographical information. The acts regulating the manufacture of cloth tell where certain kinds of cloth are made or should be made at a given date, acts prohibiting or permitting enclosure throw light on the geography of agriculture, much can be learnt of the geography of commerce and trade from acts regulating imports and exports, river navigation, canal schemes, railway concessions and so forth. Reports of Parliamentary Commissions set up from time to time to enquire into the condition of agriculture or industry or trade, are numerous and some are very useful. The journals of the House of Commons and of the House of Lords contain much material certainly of value to the geographer but as yet little worked by him.

Literary sources of great variety serve to clothe the bare bones of statistical and official documents and become more and more important as the centuries pass. Early chronicles and histories, the *Anglo-Saxon Chronicle*, or the *Ecclesiastical History* of the Venerable Bede, or the "Histories" of, for example, Henry of Huntingdon, Matthew Paris or Roger Hoveden give little geographical information. More but still not very much can be learnt from works of poetry and fiction: *Beowulf* and Chaucer's *Canterbury Tales* give a better picture of contemporary England than the chronicles but their geographical detail is scanty. Descriptive works or Topographies often masquerading under the guise of "histories" became fashionable in the sixteenth century. John Leland (1506?-1552) left voluminous notes of his antiquarian tours through England in 1534-43. William Harrison (1534-93) wrote a *Description of England* (*c.* 1577) vivid in some of its details. William Camden (1551-1623) published the first edition of his *Britannia* in 1586 and, though his interests were mainly antiquarian, the geographer gains something from his descriptions of the English counties. Local county histories began to be published about the same time: William Lambarde's *Perambulation of Kent* (1576) was one of the earliest. Some of these local histories are almost entirely concerned with the genealogies of the local families and descriptions of the "big" houses, but in many, if the reader has the taste and time to pursue the hunt, sandwiched in among the family trees and the descriptions of the furnishings of the mansions and the portraits hung there, are descriptions of the scenery, the farming, the local industries, the villages, the markets, the roads and the rivers, and the traffic along them. R. Reyce, *A Breviary of Suffolk* (1618), and R. Carew, *A Survey of Cornwall* (1602), were both written by men with an eye for country; W. Dugdale, *Antiquities of Warwickshire* (1765), and R. Plot, *The Natural History of Staffordshire* (1686) tell much about the Midlands in the eighteenth century and the iron industry developing there. Journals kept by intrepid travellers bent on business or pleasure provide, too, much good reading for the historical geographer: the *Journeys of Celia Fiennes* reflects the unusually observant eye and the very wide interests of a lady of the seventeenth century, Daniel Defoe's *A Tour Through the whole Island of Great Britain* (1724), William Cobbett's *Rural Rides* (1821-33), and the voluminous letters and journals of Arthur Young describing his travels at home and abroad cannot be missed.

Treatises and pamphlets on special topics, from those pushing daring and sometimes fantastic schemes of colonisation to those denouncing the frauds and foolishnesses of the old way and advocating the new in agriculture, industry or trade, provide ample provender, though sometimes more bulky than satisfying and always demanding the salt of critical reading. No geographer can neglect the wonderful collection of letters and journals brought together by Richard Hakluyt (1552-1616), *Principall Navigations, Voiages, and Discoveries of the English Nation.* They give an unrivalled picture of the geographical knowledge of the day; incidentally they provide some of the best reading in the English language. An " export " policy is not new and Thomas Mun, *England's Treasure by Forreign Trade* (1646) (reprinted for the Economic History Society 1933), tells " the particular ways and means to encrease the exportation of our commodities " and insists that " we must not only regard out own superfluities, but also we must consider our neighbours necessities "; the anonymous author of *The Libelle of Englysche Polyceye* (ed. G. Warner, Oxford 1926) gives in rhyme a most entertaining survey of the geography of English trade in the fifteenth century from the solid necessaries,

> " . . . the commodities
> That cometh fro Pruse [Prussia] in too manere degrees "

to the . . .

> " Apes and japes and marmusettes taylede [tailed]
> Nifles and trifles that littel availed,"

on which, according to the author, we wasted our substance in buying from Italy. A. Yarranton, *England's Improvement by Sea and Land* (1677-81) advocated increased interest in commerce. Defoe also published *A Plan of the English Commerce* (1728), and a description of the internal trade of Britain in the eighteenth century is embodied in his *The Complete English Tradesman.* Information about the techniques employed in industry and, more important to geographers, about the places where they were carried on can be found in many pamphlets: Leake, *Treatise on the Cloth Industry* (1577) (S. P. D. Eliz. Vol. CXL. No 38), tells much of the practices and more of the malpractices of the textile industry in his day, S. Sturtevant, *Metallica* (1612) and a Frenchman, G. Jars, in his *Voyages Métallurgiques* (1774) throw light on the development and location of the iron industry. Treatises on agriculture are particularly numerous, ranging from the thirteenth century work of Walter of Henley, to the comprehensive surveys of the state of agriculture in England at the end of the eighteenth century to be found in the reports issued by the newly established Board of Agriculture, one for each county, entitled *General View of the Agriculture of . . .*, and the up-to-date articles on farming practice to be found in the volumes of the *Annals of Agriculture*. Fitzherbert's *Boke of Husbondrye* published in 1523 is deservedly a classic; Tusser's *Five Hundred Points of Good Husbandry, united to as many of Good Huswifery* published in 1573 deserves mention for its engaging title and its catching doggerel as well as for " it's cool collected sense ". The practices advocated in these two treatises changed little until the eighteenth century, when Jethro Tull's *Horse Hoeing Husbandry* (1733) did much to make popular more modern methods. By the middle of the nineteenth century when J. Caird wrote *English Agriculture* 1850-1851 the major lineaments of present-day agricultural geography are already clear.

Road Books, Maps and Atlases

Road books, maps and atlases of contemporary date are of course another source of first importance to the historical geographer.

Road books and itineraries have a long history. They were popular among the travelled Romans, the Antonine Itinerary and the Peutinger Table survive as illustrations. A few medieval itineraries of travelling prelates and merchants, sometimes giving no more than a list of places, sometimes accompanied by comment, sometimes in diagrammatic form, give useful evidence of actual routes. Archbishop Sigeric's itinerary printed in Stubbs' *Memorials of St. Dunstan* gives the stages of his journey in 990 from Rome to Canterbury. Albert of Stade's itinerary of journey in 1236 from Stade to Rome gives various routes and lively comment. It is published in Latin in *Monumenta Germanica Scriptores*, Vol. XVI. Matthew Paris, a monk of St. Albans, drew about 1250 a most decorative strip map of an itinerary from England to Rome which turns at the end into a more realistic diagram of southern Italy. With these itineraries is a map of Palestine showing its trading ports and the holy places, with trains of laden donkeys and camels bringing in goods from the great caravan cities of the desert. The Matthew Paris itineraries have not been published in full, two manuscript versions can be seen in the British Museum and another in the library of Corpus Christi College, Cambridge. These are, however, exceptional itineraries and are mentioned to show that the need for road books and the idea of producing them is not new. The itineraries provided to motorists by the A.A. and R.A.C. have a long tradition and a conservative design ; in general form they do not differ much from that of Matthew Paris ; though the smudged roneograph compares very sadly with the illuminated manuscript as an art form. But the A.A. itineraries have nearer relatives. In the days of stage-coaches and before the days of ordnance survey maps there was a big demand for road books and many examples of these can be found in public libraries. The less famous and valuable but often none the less useful examples can even yet sometimes be picked up for not too great a sum in a secondhand book shop. John Norden published his tables of road distances for England and Wales entitled *Intended Guide for English Travellers* in 1625. The first measured survey of the English roads was undertaken by John Ogilby who published his results in a large volume *Britannia* in 1675. *Britannia* contained strip maps showing the roads and the details in their immediate neighbourhood and also a descriptive text. This thick folio volume measuring 11 × 17 inches was not a very acceptable travelling companion, and in the next century there are numerous small size editions of the survey more or less strictly founded on the original. The maps and an epitome of the text was issued in 1699 as *The Traveller's Guide*, being "Mr. Ogilby's Actual Survey . . . now improved, very much corrected, and made portable by John Senex". This version had many editions. *Britannia Depicta or Ogilby Improv'd* containing the maps without the text was published " not only for the direction of the Traveller [as they are] but the general use of the Gentleman and the Tradesman " by Emanuel Bowen in 1720 and this too had many editions. Another guide with a long life was William Owen's *New Book of*

Roads; the date of the first edition is uncertain, the second edition was 1779 and the last 1840. Daniel Paterson's *A New and Accurate Description of all the Direct and the Principal Cross Roads in Great Britain* first appeared in 1771 as a thin octavo volume of 77 pages but by 1829, aided by Edward Mogg, it had become a fat book of 858 pages guaranteed to stretch any pocket. John Cary published in 1798 his *New Itinerary* made "from an Actual Admeasurement made by Command of His Majesty's Postmaster General" which ran through eleven editions, and his *Travellers' Companion or a Delineation of the Turnpike Roads* in 1790. The latter had the form of a small county atlas and was very popular; it was reissued again and again, the last edition dated, as is the last edition of the *New Itinerary*, 1828. There are many more local road books, sometimes charmingly illustrated, of roads particularly frequented, roads to London and Edinburgh, to the fashionable spas, and to the Lake District, for example. The road books give information not only of the highways but to some extent of the villages and towns, their shape and size, and of the character of the countryside.

The study of contemporary maps and atlases is invaluable; it tells the historical geographer not only something of what the place *was* like at that time but also something of what it was *believed* to be like, which is often equally if not more important. For long, apart from the largely decorative mappaemundi and the severely practical sea charts or portolani, as they were called, the maps associated by tradition with the name of Ptolemy held the field. With the invention of printing, collections of these Ptolemaic maps were produced as atlases, at first little modified from their classical form; and then first the world maps and later the regional maps were altered, or substituted for, to bring the work into line with the rapidly accumulating new knowledge following upon the discovery and exploration of the ocean routes. It is most instructive to look at a Ptolemy Atlas (the British Museum, for instance has many examples): it brings a vivid realisation of the vision of our world before the sixteenth century. Early in the sixteenth century more 'modern' maps of the world appeared; one by an Italian, Contarini, in 1506, two German ones, one attributed to Waldseemüller in 1507, another to

Ruysch in 1508 are often reproduced. The cartographers of Italy, Germany and above all of the Low Countries—the leading trading nations have always a great interest in geography and hence in maps and atlases—had produced so many new maps that by the end of the sixteenth century the Ptolemy maps looked entirely archaic and the time was ripe for the publication of a completely new atlas. In 1570 Abraham Ortelius issued his *Theatrum Orbis Terrarum*, an atlas of 53 maps, a work that is often called the first modern atlas. He probably received a good deal of help from his great contemporary, Gerard Mercator, who had produced a world map on his now famous projection in 1569. Mercator too produced an atlas of his own, the first volume of which was completed and published in 1595, after his death, by his son who carried on a flourishing cartographical business in partnership with Jodocus Hondius. Willem Janszoon and Johan Blaeu, father and son, early in the seventeenth century ran another famous printing and publishing business in Amsterdam. Between 1608 and 1672 they produced a great series of atlases, the two most important perhaps were *Theatrum Orbis Terrarum sive Atlas Novus* (1635-45), volume four of which contained maps of the English counties, and the *Atlas Major* (1662) in eleven large folio volumes. Almost as famous was the rival firm of Jan Jansson, which became linked in business with that of Henricus Hondius and produced the last three volumes of the Mercator-Hondius atlas; the last volume, volume four, contained a description of Britain taken from Camden and a series of county maps from Saxton and Speed. In the next century the great world atlases were produced in France under the auspices of the Sansons (Nicolas and his sons Nicolas, Adrien and Guillaume), the Jaillots, the de Lisles (Claude and Guillaume), d'Anville and the two Robert de Vaugondys, father and son. The *Atlas Universel* of the Vaugondys appeared in 1757. To turn the pages of these great atlases one after the other in the sequence of their dates, if one is lucky enough to find several together in a good library, is a most instructive lesson.

The early English cartographers concerned themselves with the production of local and county maps rather than with world atlases. Matthew Paris' map of Britain, produced about the middle of the thirteenth century, has little detail except of the towns along the Great North Road. Four manuscript versions exist and they have been reproduced by the British Museum Trustees, *The Matthew Paris Maps of Britain* (1928). From about a century later a much more elaborate map survives by an unknown cartographer: it is in the Bodleian Library and is usually known as the Bodleian Map or the Gough Map. An uncoloured facsimile has been published by the Ordnance Survey in 1935. The first engraved map of Britain was published by George Lily in Rome in 1546. It is reproduced by the British Museum in *Six Early Printed Maps* 1928. An Englishman of unknown name drew up a map of Britain in 1564 that was printed in Mercator's atlas, and Humfrey Lhuyd of Denbigh made shortly before his death in 1568 the map of England and Wales that appears in Ortelius'. *Additamentum Theatri Orbis Terrarum* published in Antwerp 1573.

A survey of England county by county was undertaken by Christopher Saxton. He produced a beautiful series of county maps that were collectively published as an *Atlas of England and Wales* in 1579. These maps form the basis of most of the county atlases for the next two hundred years. They were published in many editions between 1575 and 1690 and still-existing copies of single counties are not rare. The whole series have been republished in facsimile in colour by the British Museum. Saxton's survey was extensively used by his more famous successor John Speed who issued his series of maps, most of them engraved by Jodocus Hondius, as *The Theatre of the Empire of Great Britain* in 1611. This atlas too had many editions: the last appearing in 1770. Speed added small-scale inset plans of the county town on his county map. This is the only complete series of early town plans, but there are many interesting early maps of particular towns. Robert Morden prepared a series of county maps still largely based indirectly on Saxton's survey, though in many Morden put in the roads from Ogilby's *Britannia*. Robert Morden's maps illustrate R. Gibson's edition, published in 1695, of Camden's *Britannia*. John Cary published his *New and Correct English Atlas* in 1787 and his county maps show a real advance on all earlier ones, both in accuracy and skill in execution.

According to Sir H. G. Fordham, Cary's plates passed eventually into the hands of Messrs. Gall and Inglis of Edinburgh, who were still using them in 1910. Cary's maps link Saxton's series to the modern maps of the Ordnance Survey.

The science of exact survey was developed first in France. The complete triangulation of France was carried out from 1744-1845 by César François Cassini and his son Jacques Dominique and the first sheet of this map *Carte Géometrique de la France* appeared about 1750. A triangulation survey of Great Britain was begun in 1784 and the Ordnance Survey was founded in 1791: the whole of the British Isles, except the mountain and barren areas which were surveyed on a scale of 6 in. to 1 mile, was surveyed on the 25-in. scale and we have now a magnificent series of detailed maps on a wide range of scales, dating from the mid-nineteenth century.

It is well worth while assembling the most complete series of maps available of the region or town or village in which the reader is interested and studying the changes the series shows. It is necessary, of course, in using early maps to try to decide how far the differences between two maps lie in differences in the knowledge of the surveyors, or in differences in the skill of the cartographers setting out their knowledge, rather than in changes in the place between the dates at which the maps were made. Some libraries have a good series of local maps: to find out what may be available, it is wise to consult T. Chubb, *The Printed Maps in the Atlases of Great Britain and Ireland* 1579-1870, and one of the various county handlists that have been compiled, e.g. Sir H. G. Fordham's "Hertfordshire Maps" published in the transactions of the *Hertfordshire Natural History Society and Field Club* 1901-07 and "Cambridgeshire Maps" published in the *Proceedings of the Cambridge Antiquarian Society* 1905-08, J. F. Curwen, "A Descriptive Catalogue of the Printed Maps of Cumberland and Westmorland" published in the *Transactions of the Cumberland and Westmorland Antiquarian and Archaeological Society N.S.XVIII* 1918, and H. Whitaker, *A Descriptive List of the Printed Maps of Cheshire* 1577-1900, Chetham Society 1942. In an area changing rapidly successive editions of the Ordnance Survey maps may be very illuminating.

There are still other local maps that have not yet been mentioned: the manorial or village surveys, the enclosure award maps, and the tithe award maps. These detailed manuscript maps often throw much light on the lay-out of fields and houses, roads and footpaths within a village and they may have, either because they are typical or exceptional examples, a much more than local importance. Some record offices have excellent collections of these maps: a good idea of the possibilities offered by this source can be gained from a study of F. G. Emmison, *A Catalogue of Maps in the Essex Record Office*. It is perhaps only fair to warn lest disappointment follow that Essex has perhaps one of the finest collections of local maps in the country: every county is not so fortunate.

III

The data culled from whatever source must be handled geographically: it must be studied in a way that will bring out the facts of distribution and of space relationship. To appreciate these facts it is always best to study *on the ground*, *in situ*, *to see for yourself*. To use maps helps in the understanding of what is seen, and maps of different kinds and on different scales will be needed.

To follow out some problems, however small the particular area under special examination may be, it is necessary to look at the local, the continental or even the world setting. A good general atlas, for example Philip's *University Atlas* or Bartholomew's *Advanced Atlas*, is an essential possession. For Great Britain Bartholomew's *Road Atlas of Great Britain* which has layer coloured maps on a scale of 5 miles to the inch is excellent; and the Ordnance Survey sheets, on a scale of 10 miles to the inch showing the distribution of rocks, rainfall, present vegetation, present land use, the population densities 1931, and the historical maps of Roman Britain, Britain in the Dark Ages, and England in the Seventeenth Century are, for particular purposes, invaluable. Incidentally, when studying large areas on maps to appreciate the distances, and the relations of part and part, great attention must be paid to the projection upon which the map is drawn: mathematicians can play very funny tricks with maps.

On the ground itself, once having set the area in its general position, the topographical maps of the Ordnance Survey are the most useful. To identify the distant parts of the view, to determine into which big river the visible stream runs, the sheets on a scale of ¼ inch to the mile are perhaps needed. For more local study the maps on a scale 1 inch to 1 mile, and if they are available, the new sheets 2½ inches to 1 mile are excellent. The 2½ in. sheets are often particularly well suited for field work since they show more detail and give more opportunity to add special features of interest to the worker than the 1 inch sheets and yet are less cumbersome than 6 inch to 1 mile sheets. For some studies 6 in. sheets are necessary and in towns perhaps even the 25 in. (or strictly speaking 25.344 in.) to 1 mile scale plans are needed. The historical geographer will often need geological maps too: these are also to be had for Britain on various scales, 10 m. to the inch, ¼ in. to the mile, 1 inch to the mile, and for consultation 6 in. to the mile. If the area cannot be visited, then maps alone must suffice for study but this is always a poor second best.

The historical geographer will wish not only to identify what he sees in the field on the map and vice versa, but often also to plot the particular set of facts in which he is interested in map form. Sometimes these facts may not be on the map—traces of old dwellings or roads, abandoned cultivation plots for example—these must then be added to 25 in., 6 in. or 2½ in. map by first surveying and then plotting them. The first volume of this series by Professor Debenham, *The Use of Geography* or Professor Debenham, *Map Making* tells how to do this. More often the historical geographer finds that modern maps are sufficiently detailed for the things that he is interested in to be fixed but they may not be sufficiently emphasised for his purpose. He may wish his map to show at a glance, for example the places names with elements of Scandinavian language, roads surveyed by Ogilby, villages making cloth mentioned by Defoe, the population density parish by parish as calculated from 1801 census figures. It is often possible to mark an existing map by underlining or colouring those features that it is desired shall stand out, choosing the map with a scale that suits the size of the area and the amount of detail required. For rough work the index sheets of the Ordnance Survey are often very useful. Those to the six inch sheets for example are published in counties on either $\frac{1}{2}$ inch or $\frac{1}{4}$ inch scale. They show all the detail of the topographical sheets in pale buff or grey with the parish boundaries strongly overprinted. The $\frac{1}{4}$ inch sheets are cheap so that one can use them lavishly in the field and study to scribble on. There is the added advantage that any special features added to the $\frac{1}{4}$ inch series can be studied in relation to the relief as shown on the topographical $\frac{1}{4}$ inch sheet and to the drift or solid geology shown on the $\frac{1}{4}$ inch geological maps.

Equipped now with maps, note book, stout boots and a stout heart the historical geographer is ready to pursue his investigations o'er fell, field and fen, down macadamed road, up cobbled street with eyes open and mind alert to see and appreciate the visible landscape as the present phase of an ever-changing pattern indissolubly linked to its past and irrevocably the foundation of its future.

CHAPTER III

THE GEOGRAPHY OF SETTLEMENT: THE PEOPLING OF THE LAND

SINCE lands without peoples are lands without history it would seem clear that the period of the first settlement of the land is the earliest period in which the historical geographer will be interested.

It might seem logical to begin with a study of the land as it was when the first settlers arrived, but there are difficulties in starting here. The physical geography of most areas then was by and large as it is now ; within strictly historical time physical changes, apart from climatic changes, have been of minor importance. A Krakatoa may disappear, a Paracutin may appear more or less over night, so may a Zuider Zee ; a dune field may change beyond recognition in a few weeks or an estuary silt up in as many decades but these are local and rare happenings. The geography of soils and of the plants and animals that the soils carried was on the other hand quite different before settlement by man. But how different ?—there's the rub. Of places settled very recently, descriptions of the look of the empty lands exist, it is true, in the letters and journals of the pioneers : these areas present no problem but they are few. Caesar and Tacitus give some idea of the look of Germany and Britain in their day : but these lands were long peopled then, and how much man had already modified them it is hard to say. The work of recent archaeologists and geographers suggest for example that the " open " belts of country and the *loess-limon* soils with which they coincide, that were formerly thought to have guided early migrants across forested Europe, may well be man-made—the result not the cause of the settlement. The geography of virgin soil is most often deduced from mere assumptions about the geographical effects of the work of man. It would not seem wise to begin here.

Man himself, groups of men, are important geographical

features. The peoples of an area have individuality, physical and cultural, that serves in a very powerful way to link or divide area and area. Within any one area whether considered one climatically, economically or politically, there is often diversity of peoples. The strength of the diversity and whether it is a diversity with or without an overriding unity, depends upon many factors; some associated with space and place, geographical, some associated with time, historical, and perhaps more associated with ideas that prevail or have prevailed, the psychological, philosophical, ideological factors. The Australian aborigines and the British colonists are just as much a part of the place Australia as the mulga scrub and the merino sheep; China is differentiated from Japan in part by the differences between Chinese and Japanese peoples. Where two or more distinct races live side by side as in South Africa or the United States, or in a different way in South America, the fact gives characteristic features to the place. The distribution pattern of French- and German-speaking peoples in Alsace or in Switzerland, of Czechs, Slovaks and Ruthenians in Czecho-Slovakia, of Jews and Arabs in Palestine are matters for geographical study. In some places it is the persistence of diversity, for example in Britain and France, in other places it is the rapidity of absorption, for example in Southern Greece and the United States, that astonishes.

Further, man himself is the most important agent of geographical change within the historic period. Although as a hunter of animals and a collector of seeds and fruits he is perhaps not much more important than any other social animal, as soon as he becomes a producer of food, a manufacturer and a trader, he may greatly modify his surroundings. Even as a family group or a tribe, a human group does more consciously to alter its surroundings than an animal group; as an organised state its power to put plans into effect increases enormously. How man works to alter his environment depends in the first place on the opportunities that the area offers to him, and in the second place on man himself, upon his ability, skill and industry, upon his traditions acquired perhaps in an area other than the one he now inhabits, upon his technological equipment, which in turn is largely a matter of contacts, place and time, upon his appraisal of the resources of the place to which he has come, and upon the length of his sojourn there.

Peoples then, for themselves as geographical features and for the work they do as geographical agents, are of first interest to the geographer. The importance of the geography of peoples needs no emphasis to a generation that has fought two world wars and attempted many peace treaties in an effort to solve the problems of the relation of peoples to land. The differences between man and man have in recent times been much emphasised for political ends. Propaganda frequently lays stress on physical differences, often where none such exist. The emphasis is placed there deliberately because a physical—i.e. a racial—difference seems more fundamental and unalterable than differences arising from upbringing and tradition. On the other hand, when it has been shown that racially two groups are alike or akin, the problems that arise from economic, social, and religious differences are not thence automatically solved. Peoples do not necessarily love their brothers any more than individuals do. The study of groups in their home may help to bring understanding and perhaps greater tolerance.

Any geographical study of any place at any time may thus well begin with a study of the peoples of the land. But in few if any areas did its people arrive all from one place at one time, nor did they necessarily arrive where they are now; the phase of settlement is a long and complex one. Periods during which settlers arrive in numbers are periods of great interest to the geographer and the study of them stimulates, in fact necessitates, interest in historical geography, for one must ask what was this place and the world to which it belonged like then.

A brief consideration of Great Britain, a small land's end peopled for the most part by the tenth century from across narrow seas, will serve as illustration.

The Peopling of Britain
The Geography of British Peoples Today

> " . . . Britain is
> A world by itself and we will nothing pay
> For wearing our own noses." CYMBELINE III. i. ii.

So Shakespeare tells us, and to some extent we do wear " our own noses " and they help to distinguish England even from France and certainly from Sierra Leone. But looks are often as much a matter of the cut of the clothes as the cut of the jib ; even superficially it is modes and manners rather than skin and bone, and more fundamentally it is way of life and thought rather than blood that distinguish the inhabitants of Britain from her continental neighbours. Their strong individuality is very largely the result of the fact that they have been *inhabitants* of Britain for a long time. Celts under Boadicea opposed the Romans, Saxons overlorded by Alfred opposed the Danes, English led by Harold opposed the Normans, but the association of people with place has been so long and so close that with Defoe it can be said :

> " Your Roman, Saxon, Danish, Norman, English
>
> . . . .
>
> Fate jumbled them together God knows how
> Whate'er they were, they're true born English now."

However, if the British abroad seem tall and florid, well groomed and stolid, at home English, Welsh and Scots are fair and dark, tall and short, lively and morose. In any one 'bus, in any one district, a great many physical types can be seen, probably more than one language heard, and certainly a great variety of opinions aired. There are clearly several strains in our stock. Observation suggests that there is also a certain regional variation in physical types and cultural groups, and statistical surveys in so far as they have been made, confirm the impression. (Fig. 2.)

Although not necessarily apparent in every pair of individuals chosen at random, there are certain physical differences, for example, between the Cornish and the Northumbrians. Among the Cornish dark hair and dark eyes are common. Many Welshmen too have dark hair and eyes, but among the Welsh there seem to be two strains ; some are moderately tall with very narrow heads like the Cornish, but another group seem markedly shorter[1] with heads of medium breadth. Short dark people of similar headform seem particularly numerous too, it has been remarked, in the industrial districts of the English Black Country and the Scottish Midlands. All down the east coast plains from Caithness and Aberdeen to Lincolnshire and East Anglia people

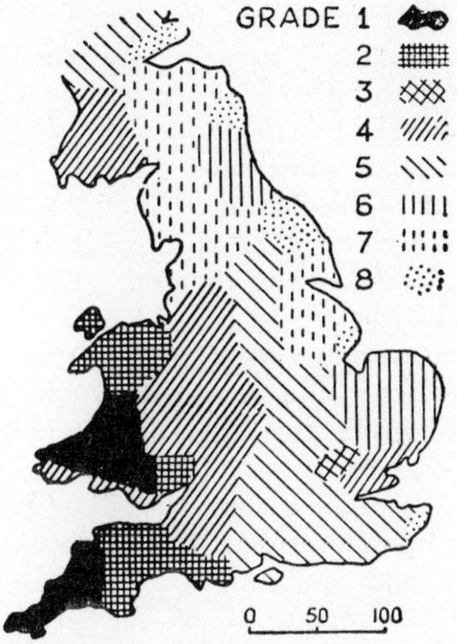

Fig. 2. Tentative Distribution of Hair and Eye Colour among Peoples of Southern Britain (from H. J. Fleure. A Natural History of Man in Britain. Collins 1951).

with fair hair and light eyes are frequently seen. Both these groups, the darker folk of the West country and the fairer folk of the eastern seaboard, differ slightly from the fair-skinned, medium-brown-haired, long- but not particularly narrow-headed, long-faced individuals who, it is said, are the commonest physical type in Britain. The thick-set " John Bull " of the cartoons with

[1] Size depends, however, more on nurture than on nature: the recent survey of the *Physique of Young Adult Males*, H.M.S.O. (1949) showed no regional pattern (not even one reflecting the depressed areas of the 30's) in the height and weight of 91,513 men between twenty and twenty-one years registered for military training under the Military Training Act 1939 and examined before the outbreak of war in Sept. 1939, except that in general countrymen were slightly bigger than townsmen.

his bigger bones and wider head though a recognisable is not in fact a predominant physical type : perhaps he is symbolic of the British character rather than the British figure.

There are clear linguistic differences between the man of Somerset, of Caernarvon and of Argyll, though all three probably listen to and can follow the news read over the wireless in "B.B.C." English. English has its strong dialects, Welsh is still widely used within the Principality, and Gaelic is still spoken in the remoter glens and islands of Western Scotland. Cornish, closely allied to Welsh, has died out only within the last few generations. There are differences in the forms of church worship : Roman Catholicism, Anglicanism, Methodism, Presbyterianism have their regional strongholds. Finally, it is not wholly looks, language and religion, still less the work of a few political cranks, that make the "English", the "Welsh" and the "Scots" conscious of their kith and kin. It is association of the groups together in the same place for a longer or shorter but here always a considerable time, and their relative isolation from neighbours that has done much to create their solidarity. It is astonishing that in an island as small as Britain with such well-developed communications a geographical pattern of peoples, physical and cultural, still exists. The physical, economic and social conditions of the present are perhaps blurring the patterns now ; but their design is still strong, the difficulties of making labour "mobile" and of enforcing evacuation of civilians from danger zones, illustrates this. In seeking an understanding, the geography of periods long past becomes a relevant, in fact a necessary, study. In Britain all the major elements of our stock arrived before the Normans ; even, according to some authorities, most of them before the Romans invaded. As far as the geography of peoples is concerned, the geography of present-day England roots in the geography of pre-historic England.

Local Sources

It is desired to find out who the peoples are who contribute to the population, where they came from, when they came and in what numbers they settled. For very recent periods there are of course records, but comprehensive records of immigration are surprisingly late and surprisingly reticent. It was only between 1851 and 1911 that the English census returns recorded the county of birth of those born in England and Wales; it was not until 1901 that the nationality of those born abroad and not British was recorded. Many other national and local records and pieces of descriptive writing refer specifically or incidentally to national and local immigration, but it is not easy even for recent periods and impossible for earlier ones to gain information that is statistically and geographically exact.

Since Britain was settled so early much use must be made of other sources. Peoples may be identified by their skeletons, by their "culture" in the archaeological sense, i.e. by the general character of their material possessions, and by their language. It must be realised that, although physical groups and cultural groups and linguistic groups can often be identified separately and independently, it is frequently very difficult to decide how these were related the one to the other, that is, to be sure which skeletons in life made and used which goods and spoke which language. Since pagan peoples usually buried a man with his personal possessions it is generally possible in pre-Christian times, if the dead were not cremated, to relate physical and cultural groups, though they are by no means always identical. Different physical types often have a common culture, a single physical type often has many cultures. It is of course impossible to link language groups securely to the physical and cultural groups until there is some documentary evidence.

The reader is advised to consider the people of his own area: to *look* at them, to observe the shape of their heads, the build of their bodies, the colour of their hair and eyes; to ponder their habits; to listen to their tongues and to say their names. His study should lead him to see whether or not his neighbours seem a homogeneous group, whether and how they differ as a group from the inhabitants of other areas he knows, and thus to proceed to ask whence and when they came to their present homes. To try to discover what groups have contributed to the peopling of a small area will give by practical experience, first understanding of the nature of sources, their uses and limitations, secondly a proof of the antiquity and complexity of the settlement, and finally an appreciation to be gained in no other way of the geographical significance of this fact.

To consider local family names is often a useful beginning. Even to run the eye over the electoral roll for county or borough will give many clues to the composition of the local population and provide much food for thought. On the shop fronts, on the war memorials, in the churchyard and in the church, family names can often be traced through the generations. Merely superficial observation will serve further to show how very local their distribution often is: the Sneezums in one village, the Gotobeds in the next; the Wallaces and the McIlwraiths lie side by side in one churchyard and but a few miles across the hills it is the McNairns and the Macmillans that neighbour each other; the Lewis' and the Jones', the Higginbothams and the Shufflebothams, the Trewarthas and the Pengellys all have their local provenance. In many places there are traces of recent immigrants: of those who "came over with the Conqueror" or of Huguenots who fled religious oppression; of Flemings enticed by Edward III or driven out by Philip II of Spain; of Dutchmen who came over to help us to drain or to farm; of Jews, Russians and Germans seeking refuge in recent decades. The occurrence for example of le, de, or van in a name—de Lisle, Deveraux, le Dieu, van Noorden, Vandelur, to select a few at random from a list of rural telephone subscribers—whether authentic and original or lately added for effect, serve to remind us of the continuous leaven immigration brings. English pronunciation does such strange things with foreign words that it is often only when consciously hunting clues that the possible original language of a name suggests itself. In most districts can be found at least the one odd and obvious "foreigner"; be the name Irish or Scots, Italian or Turk, Indian or Chinese, African or Eskimo, nothing seems too strange or too distant to appear in the most remote English village—a reminder of the fundamental geographical truth that no region is isolated, the whole earth's surface is one, especially since the reign of steel and steam. It is however only in a very few places, in Soho, or in dockside parishes of Liverpool, Glasgow or Southampton, for instance, that foreign names are numerous and varied, suggesting recent immigration on a considerable scale.

Most of the local surnames will probably show no sign of their origin, they are just all Scottish, Welsh or English, and there will be no record of the length of the residence. Now and then local memory reveals them as the deposit of a tide of holiday makers, soldiery, or " evacuees ". Just occasionally local record proves association with the district for centuries, to cite but one example, an Aspilon or Absalom paid tax in Cambridge in 1211 and " Mrs. Asplin " now hands out her buns and cakes to twentieth-century Cambridge men. Beyond the evidence of legible documents, tombstones, memorial brasses, windows and the like, it is not known what people were called.

To consider the local place names may give further help. The place names of Britain belong for the most part to five languages : French (or more exactly Norman-French), Scandinavian (Norse and Danish), English (Middle English, Early English or Anglo-Saxon), Latin, and British (Brythonic, i.e. Welsh and Cornish, and Goedelic, i.e. Erse, Gaelic and Manx). It is well worth while making a map or maps of any district distinguishing the languages of the place name elements as they are given in the *Dictionary of English Place Names* or in the local volumes of the Place Name Society Survey. Even if they prove to be all English or all Gaelic (less likely) an important fact has been learnt.

Archaeological finds will check and extend the clues afforded by the place names. To plot the recorded local finds of tools, weapons, ornaments, pottery, graves, houses and fields, forts and castles, identified by experts as belonging to particular peoples and periods, may enable differentiation between groups speaking the same language or between groups of so early a date that it is not known what language they spoke. Local finds of skeletons or bits of skeletons are less useful. First, bones are not usually found in sufficient quantity to be sure that they are typical of the area and the period. Secondly, only the very earliest invaders seem to have been distinct physically ; the dominant elements in our population cannot be distinguished one from the other by their skeletons alone.

The census returns allow a systematic study of immigration into an area in the last hundred years. From 1851 onwards the county of birth is given of those not born in the county in which they are resident on census night. There are difficulties in detail in using the figures. For example, the strength of immigration would appear very different if census night in Cambridge or St. Andrews fell in term or in vacation, in Eastbourne or Blackpool came in or out of the holiday season; but local knowledge will help in interpretation. The form of the returns themselves sometimes give rise to trouble; for instance in 1851 the enumeration is given in registration districts which make up 'resgistration counties', but the 'counties' of birth given for those born "elsewhere" are 'geographical counties' which have not quite the same boundaries. In the districts where the boundaries do not coincide there is therefore an apparent heavy migration which is in fact purely statistical. Any reader wishing to use the census returns to make studies of the local movement of population should consult H. C. Darby, "The Movement of Population to and from Cambridgeshire between 1851 and 1861." *Geographical Journal*, Vol. CI 1943, C. T. Smith, "The Movement of Population in England and Wales in 1851 and 1861." *Geographical Journal*, Vol. CXVII 1951, and H. A. Shannon, "Migration and the Growth of London." *Economic History Review*, Vol. V 1935.

The Arrival and Distribution of the Major Elements of the British Population.

It is no part of the scheme of this book to build up pictures of the geography of Britain in the successive periods during which our ancestors have invaded and settled the island. All that can be attempted here is to point out some of the constant factors that have affected the settlement that must always be remembered, and to show which are the important periods in order to suggest to the reader the periods to investigate locally and tempt him to undertake the study of their geography for himself.

Britain owes much of her character to the position of the island off the west coast of Europe and to the division of the island into a highland west and north, and a lowland south and east. The quantity and quality of her peoples and the routes by which they came are closely related to the first, the distribution of her peoples within the island to the second. The cradle of the human race is considered to lie somewhere in the stretch of country where Asia and Africa meet and Europe was thus populated from south and east. Britain lay at the end of routes from three European areas of secondary dispersal of folk. From the Mediterranean shores routes led through the Straits of Gibraltar or the Carcassonne Gap to the south-western shores of Britain; these routes would seem to have been most used by the earliest invaders during the Stone Age. From the Alpine valleys and plateaux the central route had two branches, the one via the valleys of the Saône and the Seine to the English Channel, the other via the Rhineland to the North Sea; these were the routes most used in the Bronze and Early Iron Age. From the North European plain and Scandinavia the northern route led across the North Sea with an offshoot passing round Northern Scotland through the Minch and the North Channel to the Irish Sea; thence came most of the post-Roman invaders. It is clear therefore that most immigrants will reach the southern and eastern shores of Britain; each successive wave will push some of the earlier settlers before it to the west there to join folk who have entered from the west. Thus came about the east-west pattern the British peoples still show and the marriage between the darker groups who arrived in western Britain from the Mediterranean and the Celtic speakers who came to eastern Britain from Central Europe. Sir Cyril Fox in his book, *The Personality of Britain*, makes much of the contrast in early human geography between the west and the east; the highland west is, he stresses, a region of cultural stability where new elements are largely absorbed by older ones, the lowland east a region of cultural instability where older elements are largely replaced by newer ones.

Immigrants have come in through the centuries and are still coming in, but the most important periods in the study of the geography of the British people are the prehistoric, the fifth and sixth centuries A.D., and the ninth and tenth centuries A.D.

Archaeological studies of the prehistoric period reveal that the "Britons" found by the Romans when they invaded the island were far from a simple indigenous group. Finds of pre-Roman date in Britain can be divided into groups and linked with similar assemblages on the continent: their stratigraphical sequence and geographical distribution allow the groups to be dated relatively one to the other, their centres of development established and routes of dispersal traced. The reader is referred to J. and C. Hawkes, *Prehistoric Britain*, and Graham Clark, *Pre-historic England*, for good introductions to both the manner and the matter of British archaeology.

Four ages are recognised by archaeologists before the Roman period: the Old Stone Age or Paleolithic, the New Stone Age or Neolithic, the Bronze Age and the Iron Age. These names are but convenient labels for rough divisions; neither in time nor space are the "ages" sharply distinct, flint tools continue in use far into the metal ages, Britain is still in the Stone Age when Egypt has reached the Iron Age, Bronze Age cultures persist in north-western Britain when the Early Iron is well established in south-eastern England.

The Paleolithic inhabitants of Britain, for example Swanscombe man using core flints as tools and roaming over the island in warm interglacials, and Neanderthal man using flint flake tools and enduring the cold of approaching glacials, lost the basis of their economic life with the disappearance of the big game they hunted, and would seem to have died out. The Mesolithic inhabitants, the Tardenoisians, migrating perhaps from North Africa through France and Belgium, scratching a living with their tiny tools from the sandy soils of British heaths, the Azilians from southern France collecting fish and shells and birds' eggs along the British shores, and the Maglemosians, crossing the area now occupied by the North Sea and, equipped with heavy stone axes, living in the forested river valleys of eastern England—these too, with the possible exception of the Maglemosians, contributed, it is thought, little or nothing to the stock of modern Britain, and did not do much to transform their natural environment either. The historical geographer perhaps need therefore do little more than glance at the Old Stone Age.

The Neolithic and the early Metal Age peoples on the other hand have given the island much more than a few graves and numerous tools. The immigrants of this time form a very large element in our ultimate ancestry and as the first farmers thev have deeply and for all time marked the land.

Three major Neolithic groups have been distinguished in England and each is important, for it is thought that each has contributed a distinct physical type surviving in the modern population and, what is even more astonishing, their distribution in about 2000 B.C. is reflected in the geography of the present day. A group known as the Peterborough people, with a characteristic culture of their own associated with bones that suggest that they were heavier in build than their neighbours, lived in valley and coastal sites of eastern England. They were hunters and fishers who early adopted agriculture. Physically and culturally they show links with groups who lived on the southern shores of the Baltic. C. S. Coon, *Races of Europe*, suggests that the Peterborough people may contain within their make-up elements of the Maglemosian group. He suggests further that it is back to these Peterborough people—and to the Mesolithic element in their stock—that we must look for the origin of the features that have re-emerged in the British as the type figure " John Bull ". He can see no later source for the width of the head and the heavy frame. The Peterborough folk seem to have traded in axes and tools of flint made at the workshops of Grimes Graves, Norfolk, and of stone made of augite-granophyre at the Penmaenmawr factories in North Wales. Thus they travelled far and perhaps intermarried freely with their neighbours. Furthermore, as dwellers in the forest they may have escaped in greater numbers than their Neolithic pastoralist compatriots absorption or destruction by Early Bronze Age herding groups, and as agriculturists re-emerged more strongly than they, when the art of cultivation dominated once more the economy of the island. Thus by their wide distribution, their survival and their later re-emergence they may indeed have disseminated certain of their physical features through a considerable element of the island population.

Another Neolithic group, the Windmill Hill people, so called from one of their sites near Avebury, Wiltshire, are equally important, for they too most probably have made a unique contribution to our physical ancestry. The causewayed camps that were their stockaded farms, the long barrows of their cemeteries are found scattered along the chalk downs from Devon to Sussex, and more thinly over the Lincoln and the Yorkshire Wolds and even to the Southern Uplands of Scotland. Cultural and skeletal remains similar to those found in these camps can be traced across Western France to the Mediterranean and are paralleled at various sites on the Mediterranean shores in North Africa and even in Egypt. The skeletons found suggest a small people with heads of medium width, a skeletal form associated today with dark hair and eyes and commonly found in Spain and Italy, indeed throughout Mediterranean lands. It is suggested that these invaders who reached the south-western shores of Britain about 2400 B.C. may be the ancestors of the small dark Welshmen, and of the small dark folk of the industrial districts of the west. These modern groups are seen then as the re-emergence of a surviving element of the Windmill Hill stock multiplied up by the growth of population in the nineteenth century, which was particularly rapid in the industrial areas of South Wales, the Black Country and Clydeside. The assumption seems staggering but there is no evidence of a later immigration of this Mediterranean type and no other suggestion throws any light on the present distribution of the small dark folk, with their apparent preference for remote mountain valleys in Wales, and crowded western industrial districts.

The third group of invaders in Neolithic times, the megalithic monument people, is as mysterious as the other two, and raises problems and tempts correlations just as intriguing with their suggested solutions depending on arguments as yet just as tenuous and unproven. Along the coast of Atlantic Europe remains of very distinctive tombs are found which were built apparently over a long period, *c.* 2300-1800 B.C. There are several types, but they are all mass graves. In Britain they are common along the western shores and on dry ground inland from the Bristol Channel and Wales, via Ulster, the Isle of Man, the western shores of Scotland to North-Eastern Scotland and the Orkneys, and they are even found across the North Sea in Norway and Denmark too. These tombs are also found in Brittany, north-western Spain, in

Portugal and in Sicily and Sardinia. Bones within the tombs are abundant and various, but a high proportion of the limb bones are long and the skulls unusually narrow. The distribution of these graves and this skeletal form suggests that here there is a second Mediterranean group taller and narrower in the head than the relatives of the Windmill Hill folk, who are possible ancestors for the relatively tall, dark, narrow headed Cornish folk and the taller of the Welsh folk, and they perhaps account also for the dark hair of the blue-eyed Gaels of Western Scotland. There is much controversy about the numbers of these invaders; some authorities basing their arguments on the number of graves and their wide geographical spread suggest a big migration, others see a small but powerful group of missionising overlords successfully spreading their faith, or at least popularising their burial fashions among native populations.

The distribution of finds of Neolithic date (Fig. 3) has a markedly western and south-western pattern. This may be in part because survival of their remains has been greater in the west because later invaders came in from the east. However, all the farming operations of historic folk which have always been more intensive in the south-east have not destroyed the essential south-east pattern of the Bronze Age finds (Fig. 5). It would seem therefore hard to believe that the Bronze Age groups could have destroyed the fundamental lines of the Neolithic pattern. It may be legitimate to assume with caution that the Windmill Hill and megalithic monument folk did indeed, having approached from the west and south-west, colonise the west more intensively and that this, combined with a tendency of the groups to concentrate westward before later eastern invaders, accounts for the distribution of finds belonging to them. The distribution of these finds does coincide most suggestively with the distribution of the physical elements most nearly akin to Mediterranean peoples in our population today. The correlation should be noticed if it cannot be wholly explained.

The Bronze Age (*c.* 1900-750 B.C.) is another important period in the peopling of the British Isles, for during the opening phases come the Beaker folk among whom are the last of the invaders from the Mediterranean and the only really broad or round-headed

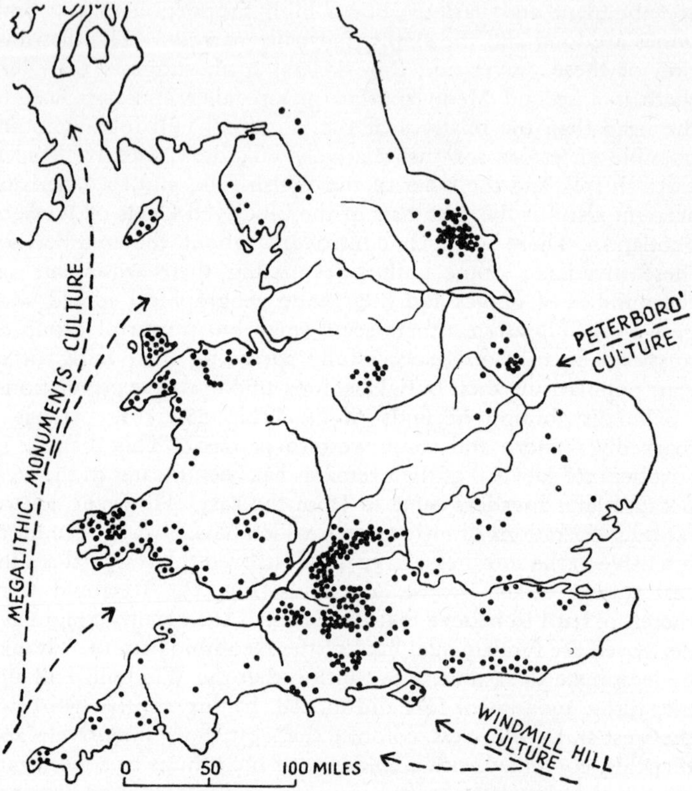

Fig. 3. *The Distribution of Finds of Neolithic Age.*
(*Based on L. Chitty and C. Fox in* " Personality of Britain ", 1943.

people to settle in Britain, and in the later phases come the first of the peoples known as " Celts ", who are held by some to be the most numerous of our immigrants relatively to the established population and thus the strongest element in our stock.

The Beaker culture, so called from the shape of its characteristic pots commonly assumed to be drinking vessels, though some

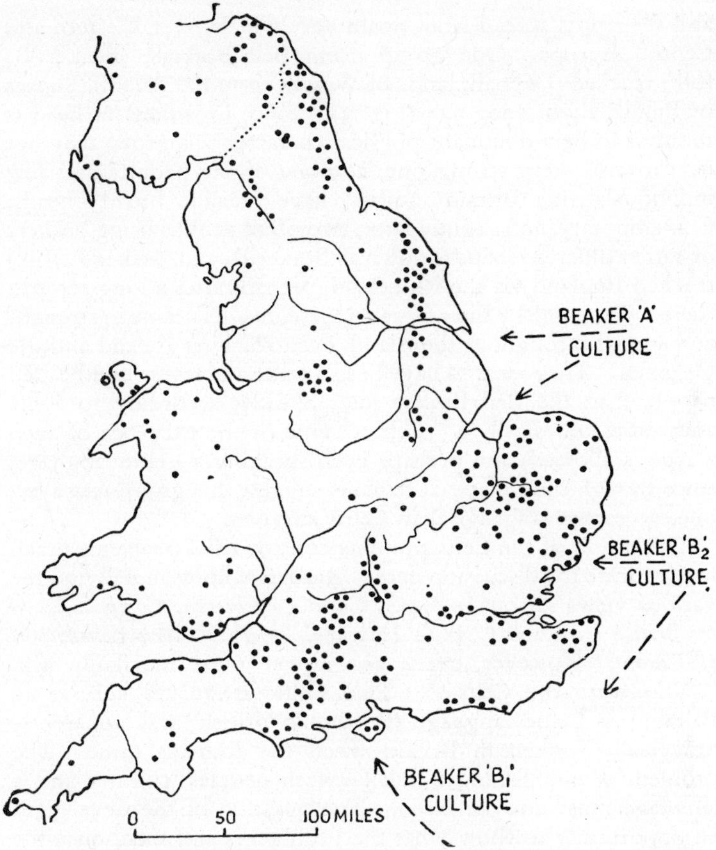

Fig. 4. The Distribution of Finds of the Beaker Cultures.
(Based on L. Chitty and C. Fox "Personality of Britain," 1943.)

would have it that they are funerary urns, seems to have originated in Central Spain. It has been suggested that the Beaker folk in their turn may have been immigrants into Spain of a mixed stock of narrow-headed Mediterranean people crossed with an older broad-headed group, perhaps of surviving Stone Age men. However this may be, a people recognised by their round heads and their pots spread from Spain very widely over Western and Central Europe. One group using Bell Beakers, (Beaker B_1 folk) reached the chalk lands of Wessex from the Atlantic shores of France about 1900 B.C. (Fig. 4). Since broadness of head is thought to be a dominant physical character this group may not have been a very strong one, as none of the peoples and few individuals in Britain today have really broad heads. Contemporary, and slightly later, two other groups using beakers of rather different shape (known as Beaker B_2 and Beaker A folk) reached England via the Rhineland, perhaps after a long sojourn there. To judge by their remains these invaders came in strength and spread throughout the island, even reaching Ireland and the Orkneys. They seem to have been a taller and less round-headed people than the Bell Beaker folk, already, according to some authorities, physically a " Celtic " type (if the existence of such a type is allowed) and perhaps even speaking a Celtic language, since though their material remains suggest this group was a big one, no earlier language than Celtic survives.

The origin of the Celts presents controversial problems much too intricate for discussion here. Readers who wish will find the various views set out in V. G. Childe, *Prehistoric Communities of the British Isles*, and C. F. C. Hawkes, *The Pre-historic Foundations of Europe*. However, every geographer must inevitably ask, " Who were the Celts ? " Two Celtic languages survive in Britain ; a Celtic language (Brythonic British) was apparently universally spoken in Britain when the Romans came. The problem of establishing a link between peoples, cultures and a language must now be faced. The question of the Celts gives an opportunity to show what the problem is and then, once the link is established, to illustrate for the first time the help that geographers can get from place names. The discussion unfortunately must be very brief and therefore far too categorical. Suffice it to state first that the distribution and character of the finds of the Middle Bronze Age suggest that following the invasions of the Beaker folk a period of assimilation and fusion of cultures within Britain ensued. In the Late Bronze Age and the Early Iron Age came renewed invasions on a broad front, not now from the Mediterranean but from north-western France and

Fig. 5. Distribution of Finds of Bronze Age.
(Based on L. Chitty and C. Fox "Personality of Britain", 1943.)

the Rhineland. The Deverel-Rimbury people, so called from two of their sites in Dorset, crossed about 800 B.C., the Early Iron Age A people bearing a culture related to that developed at Hallstatt in the Inn valley followed about 500 B.C., and the

Early Iron Age B people, the Parisi in the Yorkshire Ouse valley, the Iceni in East Anglia, the lake dwellers at Meare, Somerset, who had cultural links with the group named after their most familiar site at La Tène on Lake Neuchâtel, began to settle about 250 B.C. All these people used a light plough of Mediterranean design and this and other possessions suggest that they had profited by contact with Greek and Roman civilisation.

Secondly, several groups of skeletons associated with these cultures have been examined and it would seem that these people were of medium height and had long faces with long thin noses and low sloping foreheads, a face in fact often seen in England now. The evolution of this physical type is difficult to establish since so many peoples of the Middle Bronze Age cremated their dead. However, it is thought that they were physically as they were culturally a mixed group, emerging in the Upper Rhineland or the Alpine Lake district, dominant elements in the mixture being Beaker folk crossed perhaps with descendants of Paleolithic or Mesolithic men surviving in Central Europe and with "Nordics" of Battle Axe culture of the northern forests and eastern plains. This latter group used to be regarded as distinct from the Mediterranean race but anthropologists now tend to regard them as but another variant of it, like the megalithic monument people who migrated west, but who, on the contrary, reached eastern and northern Europe by slow migration perhaps via Asia Minor and the Black Sea or even perhaps by a wider detour across the Persian plateaux, the Turkestan steppes and the South Russian plains. These Battle Axe people would seem to be, when found in graves well within their cultural area and therefore there perhaps least mixed with other peoples, a group with a high cranial vault giving a high forehead, a characteristic perhaps associated with or linked to a trait that had a survival value. From classical descriptions these northerners may have mutated fair too. The early Iron Age invaders of Britain do not seem to have had a high vault to the head and Caesar describes the peoples of northern Gaul and Britain as reddish rather than fair in colouring so perhaps the Battle Axe element did not enter very strongly at first into the mixture of folk taking place in Central Europe. The mixture as it arrived in Britain may well not have been very different from that emerging within the island where, as has been shown, the Beaker folk also had been settling down among the earlier settlers of Mediterranean stock who had migrated west rather than east, and they had perhaps even here assimilated too a Mesolithic element from the ancestry of the Peterborough folk.

Thirdly, philologists recognise an affinity between all the languages of Europe (except Magyar and Finn) and some of those of south-west Asia, and they group them together as Indo-European languages. Wherever a European language is spoken today, archaeologists find traces of migrants bringing in the use of iron from some ill-defined area north and west of the Black Sea. Historians now begin to weigh in too. Babylonian and Egyptian records tell of invaders from the north early in the second millennium B.C., and a thousand and more years later Greek and Roman authors describe their contemporaries in barbarian lands. Herodotus writes of the nomadic tribesmen, the skilled horsemen of the Russian steppes, the Scyths, and of shepherds, the Illyrians, who kept their flocks in the mountain country between Aegean Greece and the Adriatic Sea. Cisalpine and Transalpine Gauls of the Po valley and the Rhône valley were familiar folk to the Romans. These Gauls were but the nearest of a big group : France was Gallia, our island, Britannia, was the land of the Brythons. Eastward too Gauls are recorded as raiding and settling in Macedonia and St. Paul writes his epistle to the " Galatians " in Asia Minor. Peoples are now identified by the languages they speak.

Can the cultures, the physical types, and the languages now be linked ? Was it the Battle Axe people, perhaps fused with the descendants of Neolithic farmers in the Danubian plains (for Indo-European languages seem to show a dual origin as well as an early split), who on discovering how to make iron tools and weapons were able to spread their new culture, perhaps some of their physical traits, with their language over a wide area ? Were they the parent stock of the Nasili who invaded Asia Minor about 1800 B.C., of the Aryans who swept into India in about 1400 B.C., of the Hyksos who threatened Egypt about 1100 B.C., of the Achaeans and Dorians who disturbed Aegean civilisation about 1800 B.C. and again *c.* 1100 B.C., of the Liguri and the Italici who settled in Italy? Were the Celts and the Teutons "the barbarians" of the Romans, and the Slavs, of whom classical historians tell nothing, related groups colonising to the north and west into the wilderness rather than returning south to civilisation? Perhaps so. The questions can at least be asked though answers cannot yet be given.

Certainly by classical times three groups of people can be recognised in Europe: the Hellenes and Latins of the Mediterranian, the Illyrians and Celts of the mountains and valleys of Central Europe, and the Teutons and Slavs of the northern forests. Of these three groups the Illyrians and Celts largely disappear in the course of later folk movements; they are Hellenised and Romanised in the south, they are absorbed by Teutons and Slavs in the north. In the early phases of this process, squeezed against the comparatively fully settled regions of the Mediterranean by Teutonic folk moving south to a place in the sun, the Celts migrated over wide areas east and west, colonising so effectively in western Europe that all the native place names in western Germany, northern France and Britain recorded by the earliest historians belong to the Celtic language. They colonised in sufficient numbers and had been established for a sufficient time to wipe out all signs of earlier place names if they existed, and they survived in sufficient numbers and for a sufficient time to pass on their place names—and with the names undoubtedly much else besides—to later comers. Whatever the language of the Beaker folk, the peoples represented by the cultures of the Early Iron Age almost certainly, and of the Late Bronze Age most probably, were Celtic speaking.

The last of the settlers properly called Celts were the Belgae. They are the first strictly "historic" immigrants, for Caesar tells of them in *De Bello Gallico*. According to Caesar they were quite recent settlers and they were closely linked with the peoples of Northern Gaul. Caesar's description of them and the areas they settled fits excellently with the finds of their material remains. Archaeologically the Belgae are easily recognised, they were the first group in Britain to make pots on a wheel and they used a plough fitted with a mould board to turn the sod. They appear to have crossed to Britain somewhere about 75-50 B.C. One group, the Catuvellauni, arrived in Kent and spread northward across the Thames through the forest of the Chilterns and established a capital at Wheathampstead near St. Albans and later at Camulodonum (Colchester); another group, the Atrebates, colonising further west made Silchester their capital. These capitals were towns in embryo. Belgic culture in Britain seems to be closely related to a more or less contemporary culture in the Marne valley. These cultures show contact with Teutonic groups, for the mould board plough was apparently evolved by peoples of the North German plain to till the heavy clay soil. There may even have been some intermixture with Teutonic groups, and with the Belgae may have arrived our first "English" ancestors, but if so the Teutonic element was perhaps small, for none of the place names recorded by the Romans even in Belgic areas have Teutonic elements. However, the number of Britain's place names mentioned in surviving Roman literature is small so it may be chance that none are Teutonic, but if the Belgae were Teutonic speaking it might be expected that their capital, Camulodonum, would have had a Teutonic name.

The distribution of British (Celtic) elements in the place names, although to some extent the pattern is but the negative of the pattern of the later Anglo-Saxon and Scandinavian elements, is a useful index of the extent of pre-Roman settlement. The British place names have not yet been as fully studied as the Anglo-Saxon and Scandinavian place names, and their etymology offers many difficulties because so many of the forms are only known through English sources. The English undoubtedly changed the original names very much by guessing wildly at their meaning, by mispronouncing them, and by misspelling them. To realise this it has only to be remembered what traps names like Kirkcudbright or Milngavie, let alone Culzean or Colquhoun, can be now for English tongues or pens. However, it is enough for our purposes to know that philologists can recognise certain forms as British.

By far the greater number of British place names refer to physical features, not to settlements. This may be explained in two ways. In eastern England at any rate the incoming Romans and Anglo-Saxons did not necessarily fancy the same sites as the British. They settled in new places, their new sites acquired new names and the old British village name disappeared when the old British village decayed. Furthermore, in early times when homes were not thick on the ground they had no need of identification labels, the village was *the* village, but for ready reference rivers and hills did need names. The Anglo-Saxons, like the Scandinavians in their turn, seem to have taken over well-known names for natural landmarks. All over England rivers, especially the larger rivers, have British names. Many mean just "water", e.g. Avon from *afon* water, stream; Axe, Exe, Esk, Usk from *isca*, water; Ouse from *us* water; many are descriptive, e.g. Dove, *dubo*, black; Dee, *deva*, the holy one; Taw, *taw*, the silent one; others, e.g. Frome, Kennet, Tame, Thames, are British; too but of doubtful meaning. Hill names are often British; common elements are *barro*, *brig* (Welsh *bre*, Cornish *bry*); *drum*, *penno* (Welsh *pen*). Wood names too survive; *cet* (Welsh *coed*) wood, and *rhos*, moor, heath, are common elements. Chiltern, Morfe, Blean are names of forests recognised as British though the meaning of the words is not known. It is significant that many of the territorial names are British; Kent (Cantion), Thanet, Wight, Lindsey, Deira, Bernicia are all British names. Some of the oldest towns retain a British element: London (it comes ultimately from *londo*, wild, bold), Winchester, Salisbury, Dorchester, Exeter, Cirencester, Gloucester, Worcester, Lichfield, Dunwich, Brancaster, Lincoln, Doncaster, York, Catterick, Manchester, Carlisle. In some areas in the south-west the great majority of place names, the villages and the little towns as well as the rivers and hills and the big towns, have kept their British names.

The place name evidence in general goes to show that the British were numerous and relatively well organised in all parts of Britain before the Romans came. In the west they were apparently relatively little disturbed by new and later invaders, for it is in the west that British cultural identity was not only retained but also apparently imposed on the earlier settlers from the Mediterranean region, speaking we know not what. Erse and Gaelic survive now, and elements in the place names belonging to this branch of Celtic only occur now in Ireland and in those areas of Western Britain that have received settlers from Ireland since the Bronze Age. Did the earliest Celtic speakers in England speak Goidelic and were they swept right out of Britain by later invaders speaking Brythonic or did Goidelic speakers migrate to Ireland by sea and so by-pass Britain? The latter is thought the more likely solution; otherwise the Goidelic people must have made a very short stay or the Brythonic people a very clean sweep of them and their culture, for no sign of their language to survive in the place names of England and Wales. For if they were here, they were here at or near the beginning of recorded history.

Before the Roman period, thus, all the most important ingredients of the population had already arrived; physically later settlers were to be very much "the mixture as before" with perhaps a greater element of the tall, fair people with high vaulted skulls common among the northern Europeans. Culturally, a farming economy was already established and the northern plough with its mould board was beginning to replace the light plough of the south; one of the existing languages, Celtic, was dominant throughout the island and Teutonic speakers may have begun to filter in. It cannot be denied that the geography of the pre-historic period must be of vital importance to all students of British geography.

Whatever the Romans may or may not have contributed to the geography of Britain in other ways, it seems that they contributed relatively little directly to the geography of peoples. The total numbers who came across under the Roman rule were probably small, and of these a good many of the soldiers and still more of the camp followers were people from the other side of the Channel and close relatives of those already settled in Britain. During the first phase of military activity and hostile occupation troops might be drawn by design from distant parts of the Empire and some undoubtedly, especially leaders, military and civilian, might be Romans or at least Italians. Many of those who came over, it seems, did not settle, but returned home when their term of military service ended. Of those who did settle it may well be guessed that the majority lived near the frontier towns of Chester, Carlisle, Lincoln and York, in towns like

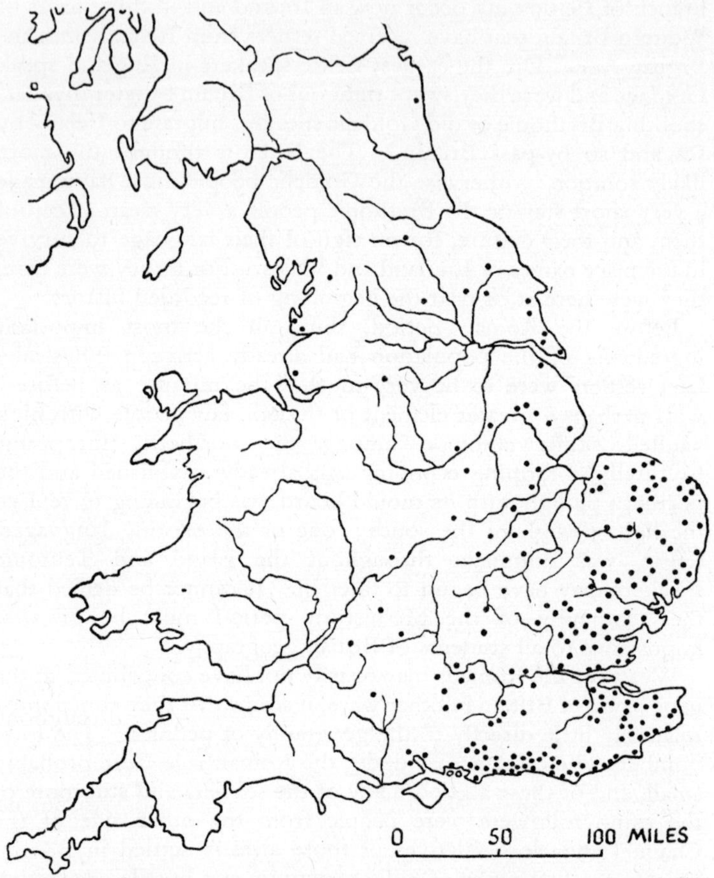

Fig. 6. Distribution of Early Anglo-Saxon Place Names: ingas *names.*

Bath, Verulamium (St. Albans) and Colchester, and most of all in London. Kent was probably the most Romanised part of rural England and a collection of skulls believed to belong to the fourteenth and fifteenth century found at Hythe show affinities with a group examined from Roman London. Both groups physically could belong to continentals, perhaps from Italy, or their descendants. Skeletons of Roman date are far too few to make secure generalisations. However, the Romans were not strong enough to establish permanently their language. There are a few Latin words surviving in the place names: *castra* gives *cẹaster*, *cester*, a fort, castle, e.g. Rochester, Wroxeter; *portus* gives *port*, a harbour, e.g. Portsmouth, Newport and *porta* gives *port*, a gate, e.g. Portgate (on the Roman wall). But these are the only common ones. Even the use of Latin in public records and legal documents is a medieval rather than a classical legacy. It would seem that the fact of a Roman invasion and occupation in the early centuries A.D. does not disturb the claim that the brunet element in our population today owes its existence not to Roman blood but more probably to Neolithic forebears who arrived in the third millennium B.C.

The Anglo-Saxons who came raiding in the third and fourth centuries and conquering and colonising in the fifth and sixth contributed, on the other hand, a very large element to the population, though how large numerically it is impossible to say. They were numerous enough to settle all over the eastern low-lands and to penetrate into the more attractive valleys of the west country, absorbing or driving out earlier groups (Fig. 6). They were numerous enough to limit later Scandinavian settlement almost entirely to regions north and east of Watling Street. They were numerous enough to introduce their tongue throughout all but the three small and isolated western districts of the island and to maintain it against the influence of the speech of all later comers. It was in fact the Anglo-Saxons who made us "English".

Written records still give no precise answers to the questions of the historical geographer: who were the Anglo-Saxons? where did they come from? when did they come? and where did they settle? Place names and archaeological finds still remain *the* sources.

The Anglo-Saxons gave most of the present names to most of the towns and villages of the island outside the Celtic-speaking areas, so place name evidence is abundant. The words used in the place names include a very large number that are common nouns and adjectives in use today and it is impossible and unnecessary to list them all. As clues to the recognition of Anglo-Saxon names two endings are more important than all the rest: *-ham* meaning homestead, farm, and *-ton* originally meaning the enclosure round the farm or even the fence but soon transferred to mean the farm itself, the village, the town. Everyone can call to mind scores of villages and towns with one or other of these endings. Sometimes the two are combined, Southampton, Oakhampton, Wolverhampton to mention three at random, and sometimes *-ham* or *-ton* is combined with *-ing* as in Birmingham, Nottingham, Bridlington, Warrington. *-Ing* is another common ending, it often stands in early names for *-ingas*, a word meaning people: Barking in early documents is Berecingas, Berica's people, Woking is Woccingas, Wocca's people and Epping is Yppingas, the people on the upland. Remembering that some names are of quite recent date, Peacehaven for example, and may be selected by analogy from any language, even the most cursory glance through an atlas of Britain will make abundantly clear the widespread character of the Anglo-Saxon colonisation. A closer analysis of almost any area in south-eastern England will reveal that place names of settlements, as distinct from natural features, are almost exclusively "English", and a study of the so-called "Celtic" areas of the south-west, e.g. Cornwall, or the Scandinavian areas of the north-west, e.g. Westmorland will reveal a surprisingly large number of English names even there.

Tools, weapons and ornaments serve not only to distinguish the Anglo-Saxons from earlier and later settlers but also to show that the Anglo-Saxons were a closely related but not a homogeneous group. All the groups seem to have been farmers equipped with the mould board plough and the differences between Jutes, Angles and Saxons, between East Saxons and West Saxons, some of which may have existed before immigration and some developed after it, are not of much concern to the geographer. The reader wishing to pursue the subject is referred to E. T. Leeds, *The Archaeology of the Anglo-Saxon Settlements*, and R. G. Collingwood and J. N. L. Myers, *Roman Britain and the English Settlement*, and their bibliographies.

The place names and the archaeological finds suggest links with continental neighbours extremely complicated in detail, but in general it may be said that the groups who sent so many emigrants to Britain in the fifth and sixth centuries seem to have been settled in the wooded lowland stretching from the Somme to the Weser and the neck of the Danish peninsula. Physically, judging from skeletons associated with grave goods in the pagan period and from written descriptions of the " Teutons ", these peoples were perhaps taller and bigger than the Early Iron Age invaders and their heads, though long, were broader with a higher cranial vault. They had perhaps a greater element than the Celt of the strain that used to be called " Nordic " but is now more often called " Northern " or " Battle Axe ". The form is very common today in Britain though many variations exist. The place names, as has been shown, suggest that both the Celts and the Anglo-Saxons settled widely and in numbers and that the Anglo-Saxons, even in the east where they early became politically dominant, absorbed many, rather than wiped out all, preceding settlers. In a grave in Berkshire of early Anglo-Saxon date eight skeletons were full grown males, all of the physical type thought to be characteristic of the Anglo-Saxons ; twelve of the skeletons were women, one only of these was of the same physical type as the men, the rest were rounder headed, smaller boned ; and among eight skeletons of children and adolescents some showed a blend of the two types. One grave may by chance preserve a record of the exception rather than the rule, but intermarriage between immigrant men and native women is at least probable.

Of the coming of the Northmen there are written records; in particular the story is told in the Anglo-Saxon chronicle. There are still some archaeological monuments but now Christianity rules the land ; " from the fury of the Northmen good Lord deliver us " pray the " English " peasants in their little churches, pagan graves with their revealing grave goods are gone. The written records as yet and the archaeological finds from now on yield little evidence of geographical interest. The place names on the other hand reveal much.

Names with Scandinavian elements in them vary from region to region not only in number but also in kind, and these variations

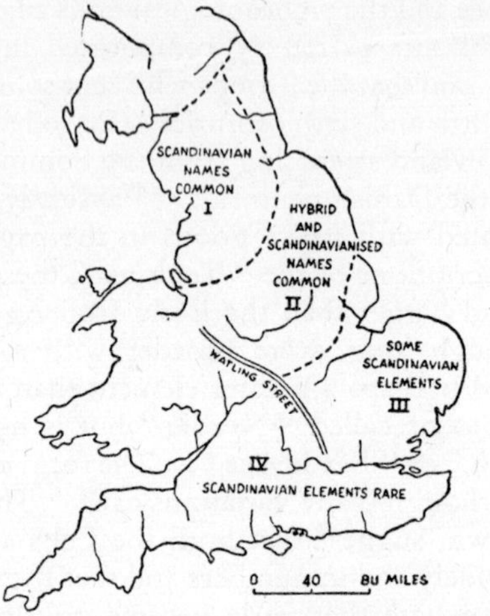

Fig. 7. The Distribution of Scandinavian Elements in Place Names.

Fig. 8. The Danelaw and the Five Boroughs.

suggest not only the areas in which the Scandinavians colonised but also the relative strength in which they colonised (Figs. 7 and 8).

In the Lake District and northern Lancashire, in parts of the North Pennine upland, in the remoter Yorkshire dales and moorlands, and also in the marshy areas around the Humber, Scandinavian names are very numerous. Norwegian names are most frequent in the Lake District; Danish names most frequent in Yorkshire, but both show migration across the Pennines. In these, the remoter areas of Northern England, the *-bys*, *-thwaites*, *-thorpes*, *-gills*, *-slacks*, *-becks*, *-tarns*, *-fells*, cover the map. Further, in these areas most of the names are wholly Scandinavian, i.e. both elements belong to the Scandinavian tongue : Grimsby, the farm or village of Grim (the Scandinavian name of the god Odin), Thurlby, Scunthorpe ; some even have the Scandinavian genitive inflexion " ar "—Amounderness (headland of Agmundr), Holderness, Amotherby. These names are clearly given by Scandinavian speakers. The names of physical features as well as of towns and villages are often Scandinavian, e.g. Scafell, Skiddaw, Windermere, and where this is so the Scandinavians may well have been the first settlers in the area. In the bigger valleys, Ennerdale, Nidderdale, Teesdale, Wharfedale, earlier peoples named the rivers, the Scandinavians took over the old name : Tees is a British word meaning heat, sunshine, Wharfe a British word meaning winding, the Scandinavians added their word dale meaning valley. In Ennerdale and Nidderdale they have added their genitive ending " ar " too.

Along the coastal plains of Cumberland and Lancashire, in the coastal districts of Yorkshire and the Vale of York, in Lindsey and Kesteven, in Rutland, Northern Northamptonshire and Leicester, in fact in a great arc around the Lake District and the Pennines, Scandinavian names are again common. Wholly Scandinavian names are frequent, but two facts suggest that, though the Scandinavians were here numerous and ultimately perhaps the dominant settlers, they were not the first settlers. The river names, Ouse, Don, Trent, Welland, Nene are British ; and many of the village names are names that the Scandinavians have found and altered rather than given in the first place, words said to be "Scandinavianised". The English (by which is meant the peoples speaking Anglo-Saxon) used soft sounds where the Scandinavians used hard ones: sh, ch, ç, d in place names the Scandinavians replaced by sk, k, t, Shipton (*sceap* [sheep] farm) becomes Skipton, Childwick (literally the children's *wic*) in Hertfordshire seems the same name as Kildwick in Yorkshire and Meathop (the middle enclosure) in Westmorland is spelt Midhop and then Mithehop in twelfth-century documents. It is common to find not only letters changed but words: the English *cirice* (church) is replaced by *kirkia*, Churcham becomes Kirkham; *stan* (stone) is replaced by *steinn* in Stainmore, Stainley, Stainton; *-bridge* or *-ridge* is replaced by *-brigg* or *-rigg*, and even *-by* is substituted for *-burgh* in Rugby and Naseby. Where such names are common it suggests that Scandinavian settlers though not the first or even the only surviving settlers, were sufficiently numerous and/or strong in the district to fix the names for ever in their form.

Further south, in an outer arc again, in Norfolk and Suffolk, Cambridgeshire and Huntingdon, Nottingham and Derbyshire, Scandinavian place names are less common and once more rather different in form. They suggest settlement in and among the English, a filling in of gaps rather than a first colonisation or a replacement and a dominance of earlier groups. Wholly Scandinavian names for first settlements are now rare except in obviously difficult areas. There are some *-by* endings in the marshy districts about the mouth of the Waveney, Ashby, Barnby, and in the Broads district, Filby, Thrigby, Stokesby, Ormesby. In other parts of these counties *-thorpes* are much commoner than *-bys* and thorpe is rarely used originally for an independent settlement but commonly for a daughter settlement attached to a nearby village. Burnham and Burnham Thorpe, Norwich and Thorpe by Norwich and many of the numerous thorpes of Nottinghamshire suggest Danish settlement alongside and perhaps at first dependent upon a nearby village forming a kind of foreign suburb. In general Scandinavian immigration here took place not during the first settlement of the area but during the later phase of expansion and the secondary colonisation of the waste.

Along the outer periphery of the last region, near the Watling Street frontier of the Danelaw for example, Scandinavian elements in place names occur, but in place names apparently given by the English. Grimston, the farm of Grim is an English word: similar forms are Thurston and Croxton. They imply a scatter of Scandinavian settlers in a district primarily English, just as in Derbyshire Ingleby, the *-by* of the English, implies Scandinavian speakers giving a name to a settlement of an isolated group of Englishmen.

Down the coast of Wales and in Pembroke, the place names also show some Scandinavian, perhaps largely Norse, influence but for the most part here the Scandinavian names seem to refer to coastal features, names that sailors might give, rather than to fields and villages as would be natural for settlers. Yet some Scandinavian settlement there must have been, for their names for the features of special interest to them to survive at all.

The Scandinavian place names have been discussed in some detail, largely as an example of the use of place name sources, just as the discussion of prehistoric settlement offered an opportunity to stress the use of archaeological material. It would seem opportune to add here two points that must always be borne in mind in using place names as sources for geographical study, though here applied in a particular instance. First, much work remains to be done locally before all the information the Scandinavian place names will yield is gained; there is but a very general outline picture available yet. Secondly, arguments can be based upon positive but not on negative evidence; where Scandinavian place names are few it does not necessarily mean that there were no Scandinavian settlers. For example, Scandinavians may have been established in greater numbers in Western Scotland than present names indicate, but perhaps on this inclement coast they did not survive in sufficient strength to transmit their names to future generations.

The distribution of the Scandinavian place names, particularly of wholly Scandinavian or Scandinavianised words, does give a definite regional pattern (Fig. 7), and a regional pattern which shows a strong correlation with regions that today show the largest proportion of tall fair folk in the population (Fig. 2).

Here then in the progress of the colonisation of our island in the ninth, tenth and eleventh centuries is an important clue to one marked feature of the geography of its peoples today. But why did the Scandinavians settle where they did? The answer is partly inherent in the unchanging facts of physical geography; the invaders approached from the north-east across the North Sea, and a few around the north coast of Scotland reached by sea the western coasts. But having arrived they found other settlers before them; they extended the settlement, they founded secondary settlements within areas already settled but settled sparsely, and they penetrated into the most attractive parts of yet unsettled country, they pushed up the dales into the highlands, on to the moors and wolds, and sought dry patches in the marshlands. To understand their pattern of settlement the geography of England in the ninth and tenth century, its plant cover as it was then, the attractiveness or otherwise of its soils as they were then to farmers with the tools, knowledge, tastes and needs of that time, the distribution of the groups already settled, not merely the unchanging facts of geographical position must be studied. Furthermore, not only the geography of England but of Europe and of the then known world must be considered, for the Scandinavian settlement of Britain is but part of the expansion of the Northmen in general. Viking voyages to Iceland, to Greenland and even to America, "Normans" colonising in France, the "Rus" settling forests and steppes of the Dvina and the Dnieper, the "Varangians" trading with Byzantium, all need attention to set the British scene, to appreciate the choices made and the evaluation given to the opportunities offered by life in the land of Britain.

The next group of settlers, or perhaps they would indeed be more truly termed conquerors, the Normans, probably brought in no new physical traits. They were Northmen mixed with peoples of the Lower Seine and there had been for long immigrants in plenty from both areas. They were certainly in sufficient numbers to leave their mark on the place names, but this mark does not suggest settlement in great numbers nor does it show any particular regional pattern to suggest concentration in any particular area. Castles, manors and monasteries founded soon after the Conquest often have French names. Many of them contain *-beau* emphasising the pleasure of owners in a new acquisition. Beamish, Beauchief, Belper, Belvoir are examples. Beaumont in Essex is a Norman name replacing an English one with a very different meaning ; Fulanpettæ, the foul pit, the English had christened it. More names were altered by the Normans than given by them : many English sounds they found difficult to say ; ch was one so that the Latin *-castra*, English *ceaster*, later, chester, became *-cester* or, as the French often drop the s before the t, *-ceter*,—Gloucester, Cirencester, Exeter. We owe the name Cambridge and hence the Cam, for if there is a bridge surely it must be over the Cam, to Norman mispronunciation. The " r-n-t " in Grontebrig was difficult to them so the old form changed to Cauntebrig, to Caunbrig, and Cambridge. Norman place names and place names changed by French tongues are scattered throughout the lowlands, but they are not numerous enough to suggest that any large proportion of our ancestors came over with the Conqueror. There are perhaps most purely Norman names in the western border districts and in the Yorkshire dales but this may not indicate that most Normans settled there, but only that colonisation was perhaps going on most actively there at that moment, and these new places needed new, and liked fashionable, names.

It is hard to assess the importance of the Norman colonisation to the British peoples. It is estimated that the population in 1086 was between 1 and 1½ million, so that even a very moderate number of incomers might be a significant addition. Politically the Normans became dominant, to some extent socially too ; but in general they modified the language of the people only a little more than their physical type. French was for a time the language of the court and the government, but English remained the language of the people, and if a few Norman barons remained distinct, many Anglo-Saxons climbed up to join them, and the great majority of their camp followers forgot their ancestry and became " true born English now ".

Post-Norman settlers have certainly contributed locally to the amalgam. The contribution of the Flemings to the early woollen industry, though it has been much exaggerated, cannot be ignored, but still more important were the Flemish refugees of the sixteenth century who helped to establish the "New Draperies" in England. The strength of the Dutch immigration is seen not only in reclaimed fenlands but in many industries, brick making, shipbuilding, glass making, metal working and printing. Visible reminders of these immigrants are to be seen in some districts, East Anglia for example, in the number of "Dutch" gables and the quantity and quality of the early brickwork. The Huguenots, the refugees of the seventeenth century, have left perhaps here and there a mark on the geography of our land especially in London and those provincial districts, Kent and Cheshire for example, which took up silk weaving. The geographical effects of the immigration to England, and in particular to London, from the late seventeenth century onwards of families, most of them Jewish, then, or soon to become, leaders in the financial world opens wide fields for exploration. All these elements were quickly assimilated; they were professional men, business men and skilled artisans and they were few in number. In the later nineteenth century immigration increased: the Census in 1871 records 105,000 foreign born, in 1881 the number was 124,000; but in 1891 there were 204,000 of "foreign birth and descent" and in 1911 285,000 foreigners are recorded for England and Wales. Jews from Russian Poland were the majority and locally they remained apart; in 1911 of the 285,000 foreigners, 153,000 were in London, of these 63,000 were Russians (mostly Jews from Poland), 27,000 were German and 8,000 Austrian among whom many were Jews. Of foreigners in London, 53,000 were recorded in Stepney; and Leeds, Manchester and Liverpool had big colonies in separate Jewish quarters.

The Scots and the Irish in England sometimes form distinct communities too, if they can be called "foreign". The Irish moved in to help to build canals and later railways, to help to unload ships in the ports, to work in industry and on the land. Much of the Irish immigration was of unskilled labour: in the Bristol dock-side parishes and in the S. Wales mining districts, in Merseyside and the Lancashire cotton towns, and in Clydeside and the industrial districts of Central Scotland an "Irish" quarter and the Irish element in the folk are still clearly distinct. The Scots migration in the nineteenth century can be followed in detail because Scotland has her own census. There was a movement of peasant labour from the Highlands to unskilled jobs in the Central Valley in the early decades; there was also a marked flow of Scots across the border, of mechanics and bagmen to the Industrial North and to London, and of farmers and gardeners to Cheshire, the Vale of York and especially to the Home Counties. The Scots of Liverpool and Manchester were still numerous and socially and culturally clearly discernible in certain districts in the first decades of this century, and the London Scottish, if less distinct geographically because more widespread, have their own regiment, clubs and churches.

However, all the post-Norman settlers, whatever their importance in certain industries and professions, at certain times, and in particular places, were too few and too widely distributed to affect directly the geographical pattern of the peoples as a whole. None of them brought new physical traits sufficient to modify the mass, or imposed new cultures on their predecessors. The fundamental British stock, that with nature has created the geography of our island, and contributed much to the geography not only of Dominions and Colonies to the most distant "outposts of the Empire", but also directly and indirectly to foreign lands near and far, was blended almost completely before the Norman Conquest and even largely before the Romans came.

The geography of the peoples of Britain might seem before examination to be a simple one; we are all an odd mixture in physical type, we all speak English, and fortunately our differences are of little economic, social or political significance now. On examination this satisfactory state of affairs is seen to be not the reflection of a simple colonisation story but of the age of the settlement. The wonder is not that assimilation has gone so far but that it is not yet complete. The long association of people with land has gone a long way towards making the inhabitants of Britain a homogeneous group but the isolation and stability born of a protected situation has also worked towards "hurrying slowly". The British have not been shocked into smoothing out every difference or so shaken that all traces of the original stratifications have disappeared.

The study of the peopling of the land is the more necessary the more complex the political geography, the more diverse the different groups, the more recent the settlement. Furthermore, the study of the distribution patterns of peoples of an area cannot be carried on in isolation in space; peoples come from and go to other areas, and yet other peoples interfere with their comings and goings. To understand the peopling of Britain for example, it is necessary to consider the peopling of Europe from whence British peoples came, and of the " new " lands overseas which received the " over-spill ".

To establish the facts of the geography of peoples in Europe is extremely difficult, especially in the marginal areas where, for practical purposes, groups most need identification. In marginal areas assumptions about existing differences between population groups are often false, and these are frequently fostered by politicians for their own ends. Attempts to unravel the story of the peopling of a part of Britain will at least make the student aware of the complexities to be expected in the story of the peopling of Europe, and give some grasp of the kind of evidence that may be used to obtain objective information about it. Western Europe like Britain was an " end "; it was protected somewhat from the assault of immigrants from Arabian and African deserts, from Asian plateaux and steppes, by Spain and the Pyrenees, by Switzerland and the Alps, by Germany and her forests. It was a region slowly and steadily colonised, a region where earlier groups largely assimilated the later and, as in Britain, a population evolved which though diverse in physical type was essentially uniform in culture. Eastern Europe on the other hand has been a passage way, and the Hungarian plain, it has been said, may be likened to a waiting-room for travellers westward. If so it has been a waiting-room at a main junction. From the forests and marshes of the Dvina and the Vistula, from the steppes of southern Russia and Central Asia, from the plateaux of Asia Minor, across the passes of the Carpathians, along the valleys of the Danube, the Vardar and the Morava, Goths and Slavs, Avars and Magyars, and Turks have poured in. Each succeeding group pushed out and replaced its predecessor, though some of all groups remained forming isolated, outlying islands in the least desirable lands. The peoples of Eastern Europe tend to be at any one time relatively uniform in physical type, but extremely diverse in culture, with refugee populations clinging passionately to their own customs, language and church. The main groups have been settled in the area since the tenth century, time, it would have been expected, to make them one. Danger from outside, it might be thought, would have fostered unity against a common foe, but the frontier was a complex, an enduring, if a shifting, one. Position on the natural routeways brought in many elements to people the land; position on the frontiers has done much to keep those elements apart. Eastern Europe has thus remained a frontier region: a frontier between Teuton and Slav, between Christian and Moslem, between Europe and Asia.

To consider the peopling of America or of another "new" country serves first to make more intelligible the early migrations of folk across Europe and Britain, nearly everyone has an "emigrant" relation which gives reality to imagination; and secondly to lead to the appreciation of the fact that the comparative stability of European designs in modern times is to a large extent the reflection of the constantly moving frontiers in America, South Africa and Australia. The geography of peoples in Britain and in Europe stirs the imagination by the contemplation of the age of the pattern. Tne geography of the peoples of the United States, for example, fascinates by the scale and the intricacy of the design and the kaleidoscopic changes it undergoes. The U.S.A. is inhabited by peoples of every hue, formerly speaking almost every known language, professing almost every known faith. There are descendants of the earliest settlers, the Amerindians, though only locally in reserves are they a dominant element in the population. There are Negroes, their ancestors brought originally as slaves from the West Indies or directly from Africa; they form a major element, and one which has a social and economic significance out of proportion to its numerical strength. The white settlers are of every nation: Spaniards, French and English were *the* immigrants of Colonial days, though

Scots, Irish and Germans came too; in the 1850's the Germans, in the 1860's the Scandinavians, in the 1890's the Italians came in greatest numbers. There are Chinese and Japanese also; they are most numerous on the West coast but their numbers have been restricted by recent immigration laws. The place names to be found on a casual inspection of an atlas map within a radius of approximately fifty miles of the junction of Des Moines River and the Mississippi in the heart of the continent sum up the complexity of the peopling of the United States well enough. Here can be found: Nauvoo, Kahoka and Keokuk; Lancaster, Colchester, La Belle, La Plata and Quincey; Memphis, Carthage and Hannibal; Warsaw, Ford Madison and Canton. The rate of increase of the population is staggering: in 1770 the estimated population was about 2 million, in 1940 over 130 million; in 1851-60 total immigration reached 2½ million, Germans and Irishmen in the lead, in 1901-10 more than 8 million arrived, over 2 million from Austria-Hungary and 2 million from Italy. The number of Celts and Anglo-Saxons migrating to Britain, of Slavs and Magyars to Danubian Lands is not known, but they shrink in imagination to a trickle beside the flood of folk who found new homes in America The rapidity of the processes of assimilation, the efficiency of the melting pot that has converted men of every race and every tongue into a single nation in less than two centuries profoundly astonishes. Even to grasp in outline its essential features, it is obvious that the pattern must be looked at in series; here again the necessity for the study of historical geography needs no bush.

To examine the peoples, their comings and goings, without reference to the work they do is dull and unrealistic. The areas from which they emigrate, the routes by which they travel, the places they choose to settle, are related to the value they place on the areas they leave and the areas they select, and this in turn reflects the work they wish to do and how they wish to do it. The present pattern of the geography of peopies is the sum of the past patterns, so past patterns must be established, but to understand the past patterns their design must be examined in terms of work. To show how to do this is the task of the next three chapters: Chapters IV and V discuss in terms of work the choice of sites of villages and farms and of towns, and illustrate the stability of patterns once established; Chapters VI, VII and VIII, discuss the effects of work and illustrate the great geographical changes wrought by it.

Map Exercises

Suggestions of maps that the reader may like to make of his own area to illustrate this chapter.

1. A map to show the regional setting of the area, emphasising its continental connections and its natural links with neighbouring British areas.
2. A relief map of the area.
3. A map of the drift geology, or if possible a soil map of the area.
4. A map or maps to show the distribution of the languages of the place names, distinguishing British, Anglo-Saxon, Scandinavian and Norman names.
5. A map of archaeological finds distinguishing the different culture groups.

CHAPTER IV

THE SETTLEMENT OF THE LAND

THE ESTABLISHMENT OF FARMS AND VILLAGES

THE pattern of settlement, the distribution of villages and towns, their size and number and building materials, does much to give geographical character to an area. The snow igloos of the Eskimo, the grass huts of the African villages of the Savánnah, the narrow shady streets of Indian bazaars, the boulevards of Paris, the skyscrapers of New York, each reflect the very varied adjustments made to widely differing climatic and social conditions. But often within a region without marked physical and social contrasts, varied settlement patterns are clear. In some places there is an even scatter of villages over the land ; in others villages are found only in certain areas, along the coasts or in the river valleys, or avoiding these only on the higher ground, leaving wide uninhabited spaces in between. In some regions single farms and small hamlets are common, and larger villages and small market towns few. In other regions from horizon to horizon houses are thick upon the ground ; square mile after square mile seems one continuous town. Some settlements have a clearly marked plan, others are amorphous. There are the villages with houses arranged in a cluster around the village green dominated by manor house, church and rectory. Others have their houses strung out along a main street with not even front gardens to break the line. Others seem to have no centre ; here the church, there the village shop and post office, yet elsewhere the smithy, now more often than not become the garage, near each a few houses. Such villages often seem to end before they well begin. Towns, too, are very variable in their plan ; the chequerboard streets of old Roman centres and many new colonial towns, the rib-like plan of streets gathering all traffic to " the Bridge ", the streets radiating from the market or cathedral square, or the repetitive rows upon rows around mine or factory.

The existence of characteristic patterns of settlement is well recognised and the study of the factors that have determined the site and form of towns and villages is obviously one to which the geographer must give attention. But all too often if consideration is given only to present conditions the pattern of settlement seems to owe its origin to pure chance and to emphasise the unpredictability of the behaviour of men. Every reader considering his own district must have pondered on the most curious choices that men have made in selecting places in which to live and work. Think of the steep streets of Cornish villages and little towns like Helston, Penzance and St. Ives ; of the climbs up and down in the Devon towns of Okehampton, Tavistock, Bideford and Plymouth not to mention villages like Porlock and Clovelly. The hilly streets of so many places in the Welsh border, the Pennine dales, and Lake District valleys are not unexpected : there as in Cornwall and Devon, man perhaps had little flat land on which to build his houses. But in Southern England, in the Cotswolds and the Downlands, the villages are often perched aloft ; even if their own streets are reasonably flat the approaching road often climbs sharply. In East Anglia, popularly considered so flat, the steep streets of many places—of Lavenham and Kersey, of Wymondham and Little Walsingham, of Bury St. Edmunds and Norwich, are seen with surprise. In general the " old " parts of most towns and villages are approached not only by narrow but also by steep lanes. What is more, it is usual to climb *up* to them in lowland districts and *down* to them in highland districts. Think too of the remoteness of many villages : villages away out at the end of a sandspit, surrounded by a river loop, on offshore islands or in remote mountain glens. What took folk to settle at Hayling Island or Sheppey or Benbecula ? Goole embraced by the Humber and Warrington almost surrounded by the Mersey, even ignoring man's despoilation, would not seem at first thought attractive places to settle in. How often are the villages off the main roads ? The, to us, obvious route between town and town passes them by ; the local roads wind in and out without any apparently rational relationship to settlement or relief. How often are villages by modern standards inadequately supplied with water and built far from the most reliable stores, on the surface or underground? Not only the village sites but the lay-out of the village lands seems odd to modern eyes. The parish boundaries turn and twist and a parish may possess, or may until recently have possessed, outlying and detached patches.

If we are not just to wonder at the results of plan or no plan but to pursue objectively a geographical study, it is always necessary to remember that the stability of the settlement pattern is very great. It is true that the archaeologist's spade or the airman's camera reveal here and there fortresses, farms and villages abandoned so long ago that they are "lost" to the modern scene, but many, perhaps the majority of settlements once established tend to persist. Man seems loath to change the site of his dwellings and prefers to go to considerable trouble to make good the deficiencies of an old site rather than move to a new one that might better fit his modern needs. Settlements change and grow but the early design is but rarely wholly obliterated and this early design was often imprinted by people with ideas, needs and equipment very different from our own. They may have been equipped with spades and hoes and ox-drawn ploughs, with horses and carts and waggons, with oars and sails, but not with mechanical excavators, electrical pumps, tractors, motor cars and aeroplanes. Any attempt to understand settlement patterns with only modern machines, power supplies and constructional materials in mind is foredoomed to failure. If we are to appreciate our villages and towns, we must try to see the country with the eyes of the first settlers, and with those of the long line of heirs that link them to us.

The choice of a site to settle was sometimes made more, sometimes less, consciously in response to a balancing of needs. The choice must often have been instinctive and incidental in detail. Except when carrying out big schemes of forest clearance or marsh reclamation or establishing model factory villages or "new towns, the whole town or village was not laid out at one time with carefully considered ends in view. What is more, the first settlers in a place must often have made false starts and some-

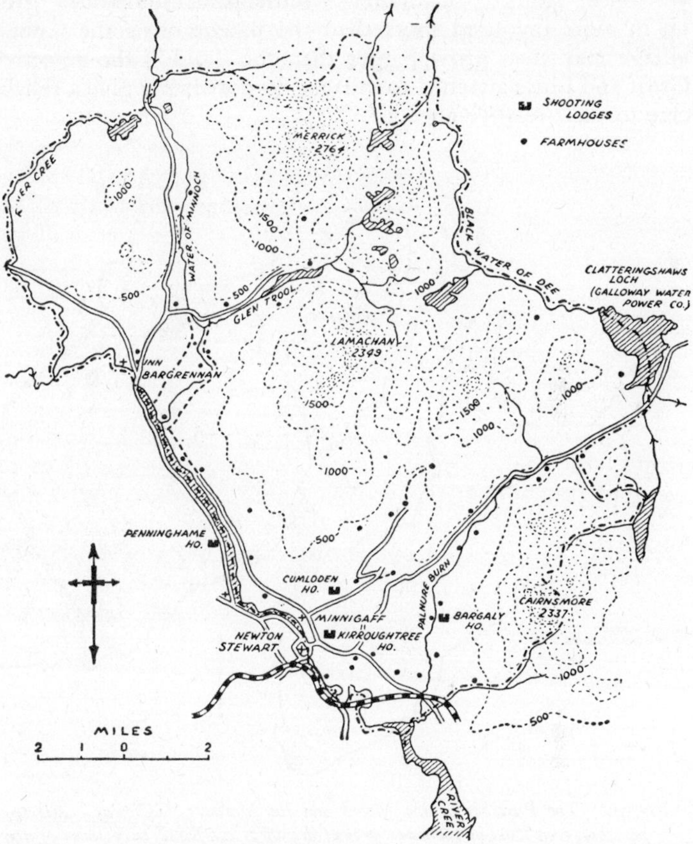

Fig. 9. The Parish of Minnigaff, Galloway. This parish illustrates the characteristic settlement pattern of a highland area. Bargrennan is the only hamlet and Newton Stewart serves as the market town for many parishes. (See Fig. 10.)

times settled down permanently on a less good site even with a better site near by, like a picnic party trying first this then that side of a bay, first one side of a hedge and then the other, and in the end putting up with much that is not ideal rather than move yet once again. "Daughter" settlements sometimes grow up in more favoured spots than the parent one; the "new" village may then grow bigger than the "old", the adjectives Great and Little attached to villages do not always give a reliable clue to their relative age.

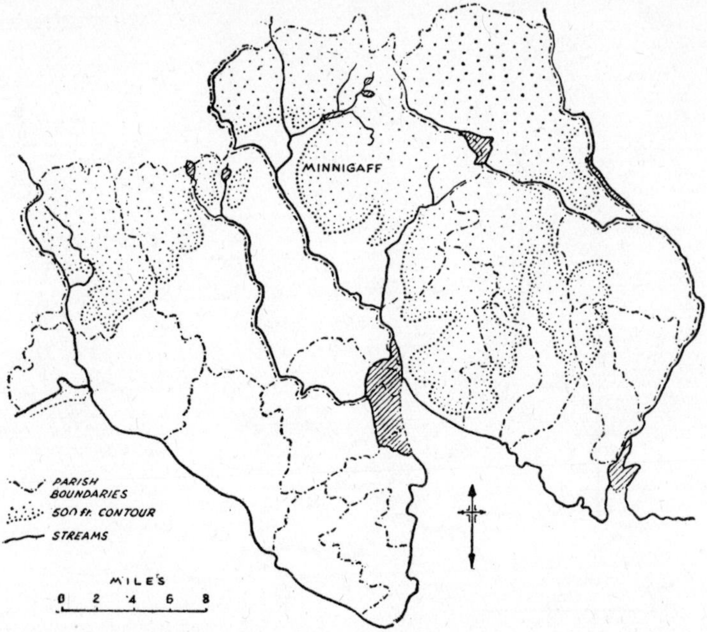

Fig. 10. The Parishes of the Moors and the Machars, Galloway. All the parishes, even those of the lower ground are large compared with those of the English lowlands.

The Villages of Britain

To illustrate these general remarks consider first the villages of Britain. They are by no means all alike; they vary in size, arrangement of streets and houses, and in the area and layout of their lands. There is in general a strong contrast to be remarked between the settlement pattern of Highland and Lowland Britain.

The Settlement Pattern

In the hill country isolated shielings in the glens, big and little farms scattered widely, each perhaps many miles from its nearest neighbour, here and there a little clachan with perhaps a school, a smithy and a church, and *the* "town" with its market and

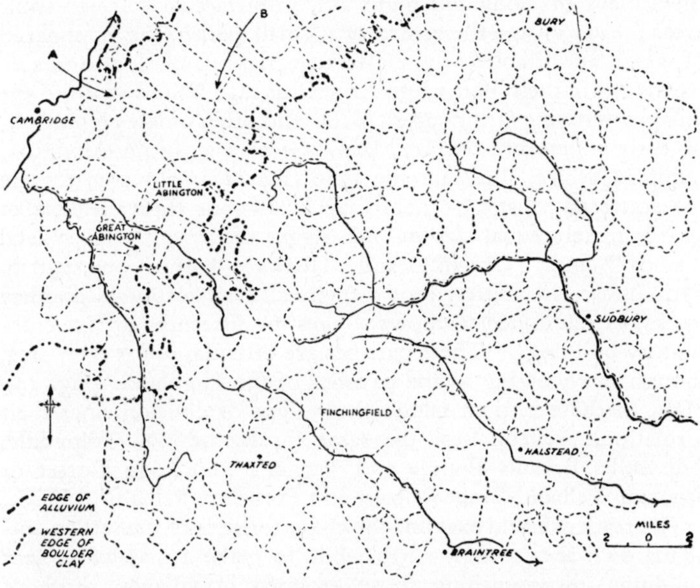

Fig. 11. *The Parishes of the Essex-Suffolk-Cambridgeshire Border and. The area shows the small parishes of the river valleys and the larger parishes of the divides. The long linear parishes* (B) *are arranged to include chalk and boulder clay or chalk and fen soils within their boundaries. A group of parishes mostly on chalk meet at* (*A*) *to share a patch of fen.*

shops serving folk living maybe twenty miles away. The farms often nestle at the foot of a hill, the clachans or hamlets are found in a valley, and the town where valleys join or open out on to plain or coast (Fig. 9). The parishes are large. The examination of a ¼ inch sheet of the Ordnance Survey or a page of Bartholomew's Road Atlas of almost any part of Scotland except the Midland Valley, of the hill country of Northumberland and

Durham, of the Lake District, of Central Wales or of the moors of Devon and Cornwall will illustrate this pattern (Fig. 10).

In lowland Britain a different design is found. Villages are in general more numerous and parishes much smaller (Fig. 11). The villages are of two kinds : both kinds may often be found in one area. Compact and linear villages are most often found in the valleys or along old highways, in attractive and accessible areas ; each village the centre of a small parish, the boundaries of which are often clearly related to variations of slope and soil. Farms with their barns and cowsheds are found within the villages, outlying farmhouses are few and often of late eighteenth- or early nineteenth-century date. Great and Little Abington, Cambridgeshire, illustrate the type (Fig. 12). In contrast are the scattered villages ; there may be two or three hamlets of approximately equal size or one bigger settlement with several " Ends " and " Greens " and " Rows " within one parish. Farm houses are scattered too ; their hall plan, elaborate chimney stacks, or mullioned windows suggesting fifteenth- or sixteenth-century buildings. These parishes are often large for their area, though rarely as big as the parishes of the Highlands (Fig. 13). Many of these in their names show signs of division since their formation ; several bear the same " surname " as Tolpuddle, Affpuddle, Bryants Puddle and Turner's Puddle in Dorset or Stocking Pelham, Brent Pelham and Furneaux Pelham in Essex.

The reader should examine the characteristics of the settlements of his own area. It is worth while to make a map of present buildings, distinguishing those grouped in villages, those of scattered hamlets and the isolated farms and cottages, and a map of the parish boundaries to show clearly the variations in their shape and size. The pattern should be studied in relation to the local relief and soils, but, before attempting to consider the factors which may throw light on the correlations noticed, an attempt should be made to determine as far as possible the date of the establishment of the settlements and their boundaries.

The Age of the Settlement : Sources

A series of simple maps using sources readily available will indicate something of the evolution of the settlement pattern

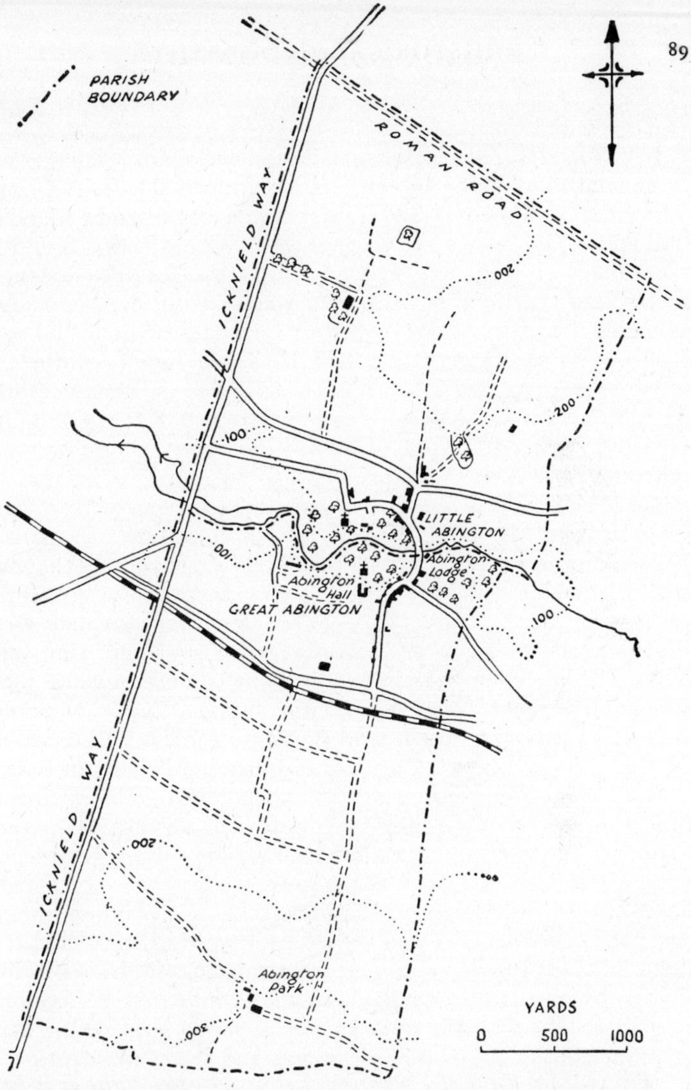

Fig. 12. The Parishes of Great and Little Abington, Cambridgeshire. These parishes show the characteristic features of closely settled river valleys; the river and the valley crests form the short boundaries, the villages cluster in the valley around church and manor: outlying farms are few. The old ads are used as parish boundaries.

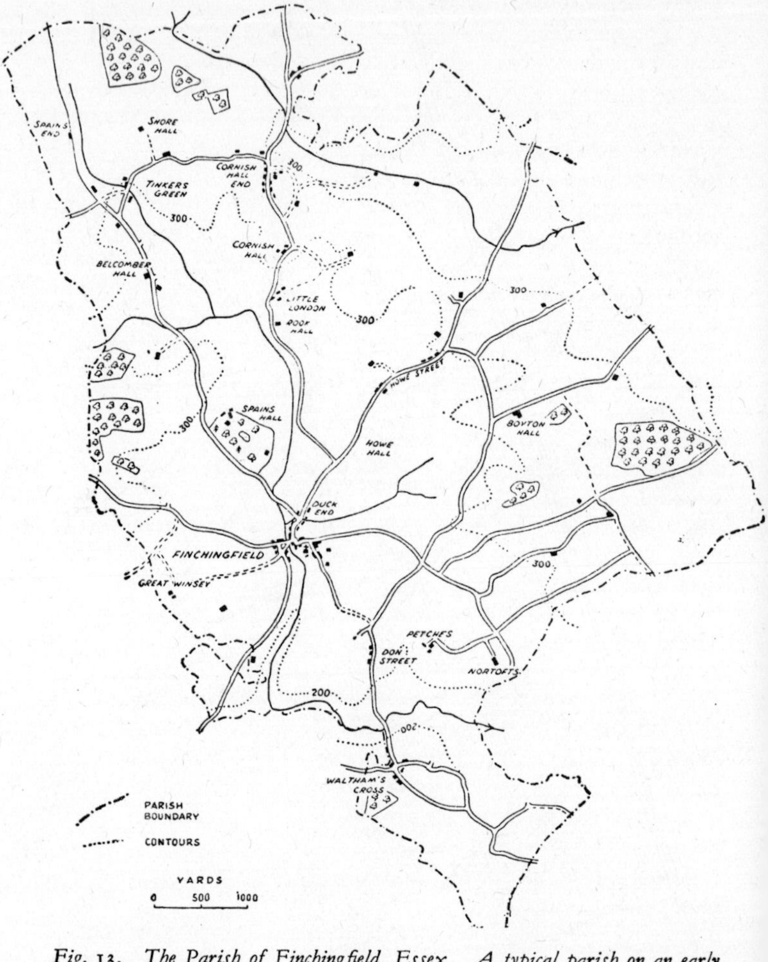

Fig. 13. The Parish of Finchingfield, Essex. A typical parish on an early frontier of settlement. The large area of the parish, and the number of hamlets and scattered farms of early date are characteristic of frontier areas.

and greatly increase understanding of it, however imperfect the material from which they are made and inadequate our use of it. The period of colonisation is a long one and the date of the first permanent settlement varies very much from area to area; colonisation in the strict sense is still going on, new land is still being taken up and new settlements formed. The north and west was in general very much later in reaching anything approaching full settlement than the south and east.

The parish boundaries in many areas have been changed in recent times. In 1801 the parish for rating purposes (i.e. the civil parish) was still for the most part coextensive with the ecclesiastical parish, for in Elizabeth's reign the existing ecclesiastical parishes had been taken as the unit for Poor Law relief. But by the middle of the nineteenth century in some places new population centres had grown up that were separately rated from the parish within which they lay, other places had grown so large that they had swallowed up neighbours once distinct, and the detached parts which many parishes possessed had become tiresome anachronisms. As a result of a series of Acts dating from 1832, where it seemed convenient for administration new parishes were created, old parishes were amalgamated, and detached parts attached, either to parishes within which they lay or with which they had the longest common boundary. These adjustments were by no means universal; the majority of rural districts remained untouched but where changes in the parish boundaries were common it is informative to map if possible the earlier pattern. Early editions of Ordnance Survey maps, information from the census returns where mention is made of changes of parish areas within the decennial period, and the maps of local authorities will supply the information needed.

The parishes of the beginning of the nineteenth century may have existed for a long time. In many areas they do not differ from the parishes listed in the Valor Ecclesiasticus of 1534-35 or even from those in Pope Nicholas' Taxation of 1291. How far these represent the original local entities of settlement remains a problem. The relationship of "parish" to township is obscure; in some old corporate towns, the

" township " included several parishes, but in sparsely populated areas one parish may have within it several townships. Both township and parish are related, though in complex fashion, to the manor. Blackstone believed that " the boundaries of the parish were originally determined by those of the manor or manors. With the spread of Christianity the lords began to build churches upon their own desmesnes or wastes . . . and obliged all their tenants to appropriate their tithes to the maintenance of the officiating minister ". Subdivision of some of the larger parishes has certainly occurred since the original organisation of the communities. The parishes with common names suggest this. The subdivision in some cases is pre-Conquest for sometimes the Domesday Survey records two or more vills of the same name. Subdivision in frontier areas probably went on steadily, and in some areas particularly in the seventeenth and eighteenth centuries, as the Act of 1662 that recognised and legalised the process bears witness. In some regions then it may be profitable to make a third parish map showing as one unit parishes which record or name suggests have been subdivided ; the large parishes so obtained may help to indicate the frontier areas of the original settlement.

Maps at significant dates of the settlements themselves are next needed. The most rewarding date to choose for each map of the series will vary from district to district but for most areas it will be useful to make at least one for a nineteenth-century date, one for a seventeenth-century date, and for southern districts, one for 1086, the date of the Domesday Survey. For northern districts a map shewing the dates of the foundation of parish churches is perhaps the best substitute for the last map. Carefully used, maps of certain place name elements may give some further clues to the progress of settlement.

A distribution map of buildings as they are shown on the earliest edition of the Ordnance Survey sheets, and in some districts for one or more dates between that and the present day will serve most emphatically to illustrate the extent of recent expansion of settlement. The next map may show the towns, villages, hamlets and mansions marked on the county maps of one of the famous cartographers Saxton, Norden or Speed ; this will show those considered of most importance in the sixteenth century. To leap the centuries again to mark the vills recorded by the Domesday Commissioners for 1086 gives in general for most areas then considered worth surveying a fairly reliable map of the settlements of the eleventh century. This map will give, especially for the areas on the then frontiers of active colonisation, a conservative picture, for the vills mentioned are those of manorial holdings separately accounted. Some outlying farms and hamlets, " wics " and " denes " in marsh and forest, geographically but not yet tenurially distinct, were included in the returns of the manor to which they belonged and thus escape record. A map of Domesday vills shows but few in the Weald for example, and therefore suggests that this area was less settled then than may have been so. In most areas of lowland Britain however, a comparison of a map of Domesday vills with one of present settlements shows a remarkable correspondence and drives home more strongly than anything else can the age of the settlement in these districts and the stability of the pattern. Some few Domesday manors have disappeared or remain only as farmhouse or field names now; some few areas even in early settled districts seem to have been colonised since the eleventh century, but they are few. To check and in some districts essentially to extend the settlement pattern given by Domesday vills, in the far north and west to establish the original pattern of settlement, a map may be made showing the earliest authenticated date of the parish church. Failing evidence of the date of the foundation of the church, the period of the earliest surviving part of the fabric may be mapped as the latest limiting date. It must always be remembered in interpreting this map that people may dwell in an area long before a church is built, and conversely that some few churches were erected " in the wilderness " and did not necessarily bring at once permanent settlement. The earliest known documentary mention of a place name (this information is given in the *Dictionary of English Place Names* and in the county volumes of the Place Name Society) will also give a latest limiting date for the naming of the place; to map these dates may help in the elucidation of the period of active colonisation in some areas of late settlement. The later the period of colonisation the less likely is it that inhabited places will long escape mention or that all documents which record it will be lost. A distribution map of places mentioned in pre-Conquest documents on the other hand is not likely to give a significant settlement pattern ; it gives but a chance selection.

The examination and mapping of the place names themselves will aid in establishing broadly between region and region, and within one region in more detail, the date and the sequence of settlement. Omitting names of places recorded as established in recent times which may have by analogy names of any type, place names can be arranged in roughly chronological sequence. The language of the place names gives some clues. In general in any one area British names are given earlier than Anglo-Saxon, and Anglo-Saxon than Scandinavian, but for example new villages in say, Cornwall may be given British names contemporary with or even later than Anglo-Saxon or even Scandinavian villages in, say, Northumberland. The map or maps made to show the elements in the local population (see Chapter III) will also then illustrate broadly the progress of the settlement. The meaning of the place names may also suggest the relative date of the naming of a place. Locally and in detail etymologists can pick out personal names and other words that belong characteristically to certain periods, as the language evolves certain vowels and letter combinations go out of use and particular names and words go out of fashion. These tests are too specialised to be applied by any but experts, but some more general considerations can be appreciated by the layman. In a local study of settlement it is well worth while to map certain groups of place names, provided that it is realised that the patterns given are useful in that they may suggest clues to aid in checking the interpretation of patterns of settlement obtained from other sources rather than in allowing a rigid interpretation based upon their evidence alone.

Places that contain the names of pagan gods, like Wednesbury (Staffs) from Woden, Tuesley (Surrey) named after Tiw or Thundersley (Essex) from Thunor were probably given before the conversion to Christianity. Names containing words connected with pagan worship like Harrow (Middlesex) from an

Anglo-Saxon word *hearg* meaning temple or Weedon (Bucks) and Willey (Surrey) which both have *weoh*, a heathen temple, as a first element must also belong to the pagan period. In south-eastern England they probably date from the sixth century or earlier though in the north and west they may be a little later. Pagan names in the north and west, are, as a matter of fact, rare, suggesting that the spread of Christianity soon caught up the spread of settlement. Place names that are folk names, i.e. names of groups of people that have been transferred to places, are probably an early form. The place names ending in *-ing* where *-ing* is not an adjectival form but stands for *-ingas* meaning people, like Hastings, Haesta's people and Reading, Reada's people, are the most important of this group but others will be found. The *-ingas* names probably belong to the first phase of the Anglo-Saxon settlement and they show a strong south-easterly distribution pattern (Fig. 6). The endings *-ham* and *-ton* especially when compounded with a personal name, e.g. Aylsham (Norfolk), Ægel's homestead, and Brighton (Sussex), the village of a gentleman named Beorhthelm, are also considered to be early forms; *-ingham* and *-ington* like Birmingham, the homestead or village of Beorma's people and Workington, the village of Weorc's people are common too. These names suggest a simpler society with fewer settlements than one which gives detailed descriptive names as for example Redmarshall (Durham), a hill by a reedy lake. Although of course a general rule cannot be strictly applied, as a group the *-ham* and *-ton* names of an area belong to the first phase of its settlement. It is considered too that, very broadly speaking, the majority of names ending in *-ham* (when they can be distinguished from those originally ending *-hamm*, a meadow) are earlier than those ending in *-ton*, for *-ton* was a very common place name element for a long time and was certainly much used even after the Norman Conquest. The endings *-ham* and *-ton* clearly imply an inhabited place; many place names, though they imply that the place was frequented, otherwise there would have had no need of a distinctive name and no chance of the name surviving, do not necessarily imply that at the time of the christening the place was permanently inhabited. Names like Brough and Chester, Oxford and

Bridgwater, Portsmouth and Whitehaven and a host like them may not as a group perhaps be as early settlements as the earliest *-hams* and *-tons*, but important forts, bridges and ports, if not original settlements of a farming community, soon attracted more inhabitants than the purely agricultural centres and early grew into towns. The name elements so far discussed, the

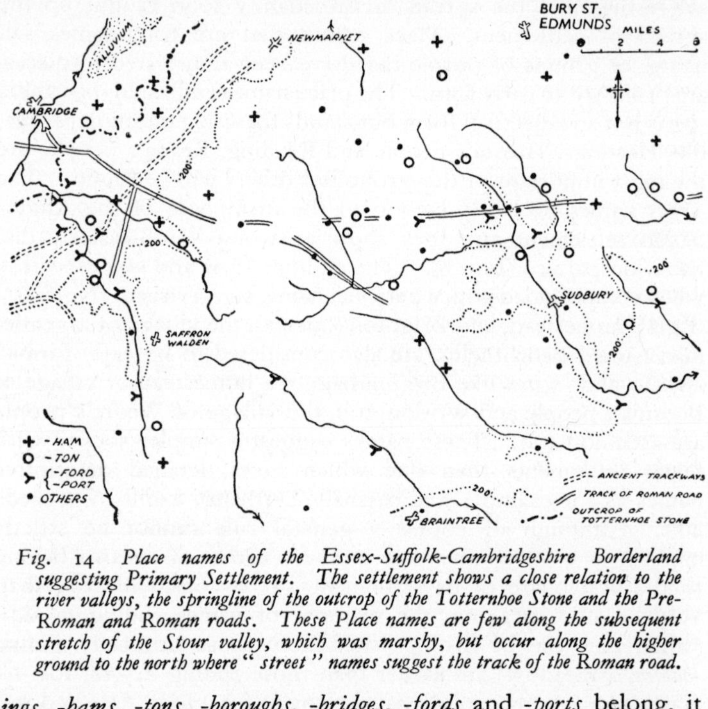

Fig. 14 *Place names of the Essex-Suffolk-Cambridgeshire Borderland suggesting Primary Settlement. The settlement shows a close relation to the river valleys, the springline of the outcrop of the Totternhoe Stone and the Pre-Roman and Roman roads. These Place names are few along the subsequent stretch of the Stour valley, which was marshy, but occur along the higher ground to the north where " street " names suggest the track of the Roman road.*

-ings, *-hams*, *-tons*, *-boroughs*, *-bridges*, *-fords* and *-ports* belong, it is thought, to places that from the first or very early were established as settlement units organised to a large extent independently one of the other. The distribution pattern shown by a map of them collectively may give an indication of the lines and areas of primary colonisation (Figs. 14, 16a, 16b, 17).

There are in addition other place name elements which occur very frequently in all areas which suggest the hiving off of later groups, founding settlements perhaps at first administratively attached to the parent settlement. Villagers, it would seem, frequently made, some distance from the village itself, small-scale clearings in the wood, small innings in the marsh or used pastures on high ground to grow extra grain or flax or feed extra horses, cattle, sheep or swine. Certain place names suggest some such origin. The elements *stead*, *stow* and *stoc* meaning essentially just "a place" are very common in many regions. *Stead* is rarely compounded with a personal name but frequently with a descriptive element: with a tree name, Buxted (Sussex), Elmstead (Essex); with a land name, Felsted (Essex); with a crop name Linstead (Suffolk), Plumstead (Kent); with beasts Horstead (Norfolk), Swinstead (Lincs). *Stow*, alone or with a personal name, often meant a holy place, a meeting place for worship. It is frequently found with a saint's name: Stow-on-the-Wold (Glos) occurs in some early documents as Stowe Sancti Edwardi; similar in formation are Edwinstowe (Notts), Felixstowe (Suffolk) and Godstow (Oxon). It is not unlikely that these were sometimes remote from the village to which they belonged in a secluded spot to give or preserve aloofness from common affairs. Chepstow (Mon) the market place, Plaistow (Derby), the play place, and Burstow (Surrey), the place by the burg, suggest other meeting places. *Stoc* was also used for a meeting place, sometimes with the special sense of monastery, but the great number of "Stokes" suggest that places so named were common adjuncts to many villages. These were later distinguished one from the other, as increased ease of travel made it necessary, by adding, usually in post-Conquest times, the name of the village to which the *stoc* was originally attached: Basingstoke near Basing (Hants), Chardstock near Chard (Devon), Stoke Newington (Middx), Stoke upon Trent (Staff), Stoke Bardolph (Notts) are examples. *Thorpe* is in general the Danish equivalent of the English Stoke; Thorpe does occur as an Anglo-Saxon word meaning village, usually a smaller one than *-ton*, but in English place names the element is far more frequently a Danish word meaning "a village due to colonisation from a bigger one". Thorpe, like Stoke, sometimes stands alone as a name, but since there were so many

Thorpes most have acquired an additional name. Scunthorpe and Mablethorpe (Lincs) are without mother villages of the same name; Cleethorpes (Lincs) is now an enormous child of its little parent Clee; but Barkby and Barkby Thorpe (Leics) and Burnham and Burnham Thorpe (Norfolk) still illustrate the original relationship. The Anglo-Saxon *-cot* and *-wic* and the Scandinavian element *-booth* are three other elements which, though emphasising the huts not the place, probably were usually given to outlying settlements. *Cot* applied to any building but very frequently it would seem to be a shelter for animals: Sapcote (Leics) is *sceap-cot*, a shelter for sheep; sometimes *cot* is grandly changed to *-court* as in Maidencourt (Berks). *Wic*, a dwelling, is often found with words that suggest it meant dwelling on a farm. In the form *bere-wic*—Berwick (Northumb), Barwick (Norfolk), Berrick (Oxon)—literally a barley or corn farm, the word acquired a semi-legal status. It is used in the Domesday Book to describe an outlying holding belonging to a manor. The element is frequently compounded with an animal name: Shapwick (Dorset), Hardwick (Cambs) the sheep farms; Gatwick (Surrey) the goat farm; and most commonly of all it is compounded to suggest a dairy farm—Cowick (Yorks) cow farm, Cheswick (Northumb), Chiswick (Middx and Essex), Keswick (Cumb) cheese farms, and Butterwick (Durham, Yorks, Lincs). When the word occurs alone as in Week (Cornwall, Hants, Wilts), Wick (Som, Glos, Worcs, Berks), Wicken (Cambs), Wix (Essex) and Wyke (Dorset, Surrey), Ekwall in his *Dictionary of English Place Names* gives the most probable meaning as dairy farm. *Wic* in one form or another occurs in names of many places that have an importance other than agricultural. It would seem to be especially associated with salt workings as in Nantwich, Northwich, Droitwich; indeed in the Domesday Book " wic " is often used for a saltwork. It is frequently found in south-eastern England in names of ports, Norwich, Dunwich, Ipswich, Harwich, Greenwich, Sandwich. It is possible that here the element had originally no special significance; saltpans and saltmines usually belonged initially to nearby village or villages, buildings would grow up around them and the name at first referred no doubt to the huts not the workings. Similarly at the nearest point on navigable water to a village, a hithe and then dwellings or storehouses might be built; a little outport grew and the huts might be emphasised in the name as in Greenwich, rather than the quays as in Rotherhithe. The wide market for salt, the far-flung traffic of the estuaries often resulted in the original outpost persisting, the

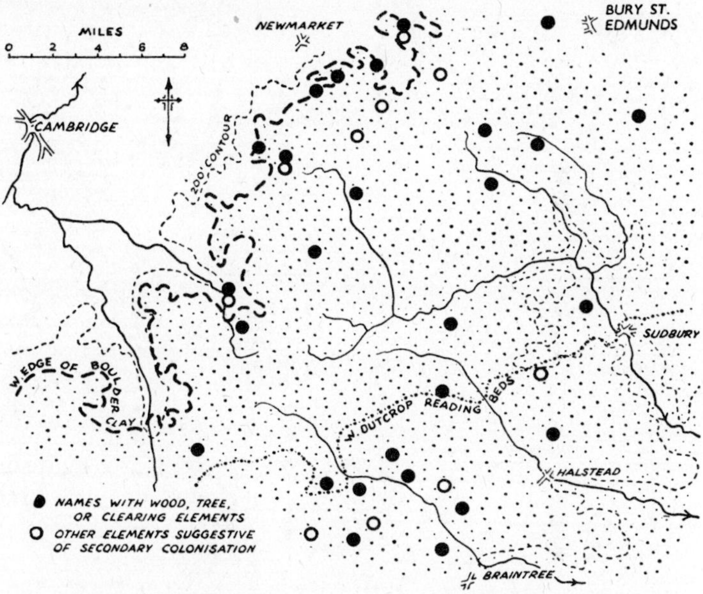

Fig. 15. Place names of the Essex-Suffolk-Cambridgeshire Borderland suggesting the expansion of settlement from the primary centres.

original village to which it belonged disappearing. A great many place names contain elements that suggest at the time of their christening a place in a wood; they are very numerous in all areas, many now belonging to big villages and even towns, which may well have had an origin as little clearings in the wood made by villagers already settled on its edge. These names will be discussed in Chapter VI; the endings *-field* and *-ley*, meaning land free from wood, are perhaps most common.

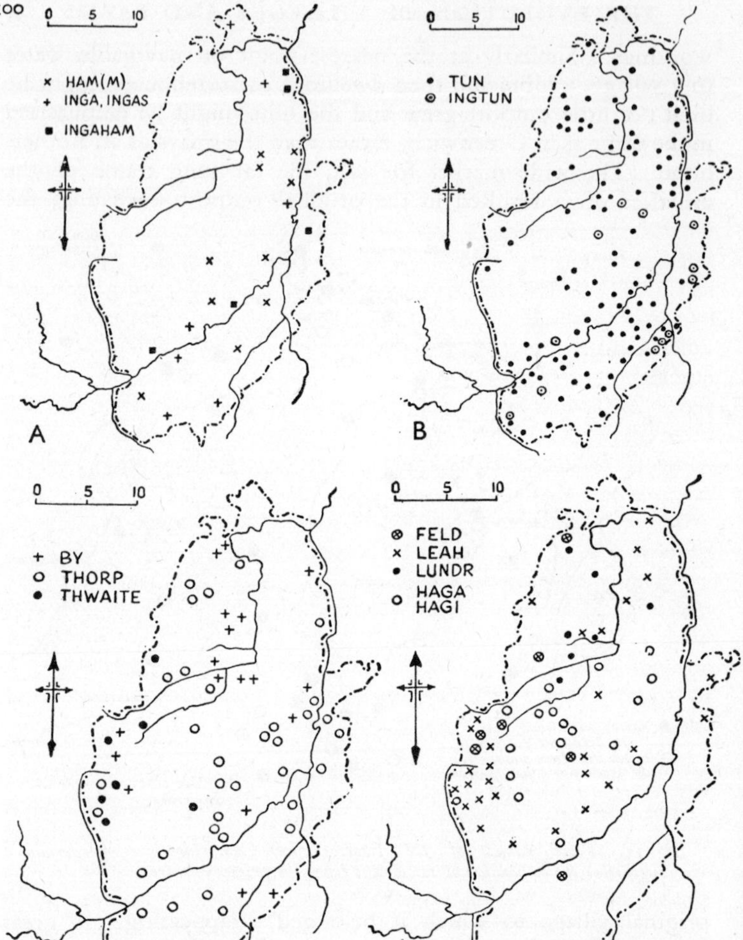

Fig. 16. *Place names of Nottinghamshire suggesting progress of the settlement.*
(*A*). *Early colonisation along the rivers.*
(*B*). *Extension of primary settlement.*
(*C*). *Scandinavian names suggesting further extension of the primary settlement* (-by *names*), *and the beginnings of secondary settlement* (-thorpe *and* -thwaite *names*).
(*D*). *The colonisation of the woodland.*
(*From J. E. B. Gover, A Mawer, and F. M. Stenton,* "Place Names of Nottinghamshire", *English Place Name Society. Vol. XVII.*)

A map of these elements, the *steads* and *stokes*, the *cotes* and *wics*, the *fields* and *leys* and numerous others of general and local occurrence which have a similar significance, may suggest something of the spread of secondary colonisation from the original centres (Figs. 15, 16c, 16d, 17). A comparison of the distribution pattern made by these elements with that of names which suggest primary settlements may suggest the progress of settlement within a region. It must always be remembered that secondary settlement in an accessible area may long precede primary settlement in a more remote region, but towns and villages do reveal in their names something of the date and circumstances of their origin, and for the analysis of the earliest settlement patterns in many areas place name elements, for all their uncertainties, may provide the only clues there are. Place names for geographical study have the advantage that, unlike archaeological finds and surviving documentary records, they provide source material well and widely distributed throughout the area.

When by a study of the archaeology and architecture, of documents and place names, an attempt has been made to find out what is known about the age and progress of the settlement in an area, then comes the even more difficult task of attempting to appreciate the factors which have determined the choice of the sites of the villages and towns, and the lay out of their lands.

The Factors Important in Determining the Pattern of Rural Settlement

The evaluation of the opportunities afforded by the virgin land through which the settlers spread was governed by the need of dry and level land on which to build, of a reliable water supply easily reached, of good pastures and fertile soil easily husbanded, of ready access from landings, of contact with friends, and of shelter from foes. Highland and Lowland Britain differ in the opportunities they offer for the satisfaction of these needs and therein in part lies the explanation of the different settlement patterns observed in the two areas. Highland Britain has a high rainfall, strong winds but an equable climate ; Lowland Britain is drier and summer temperatures are higher. In Highland Britain rocks geologically old outcrop ; they give

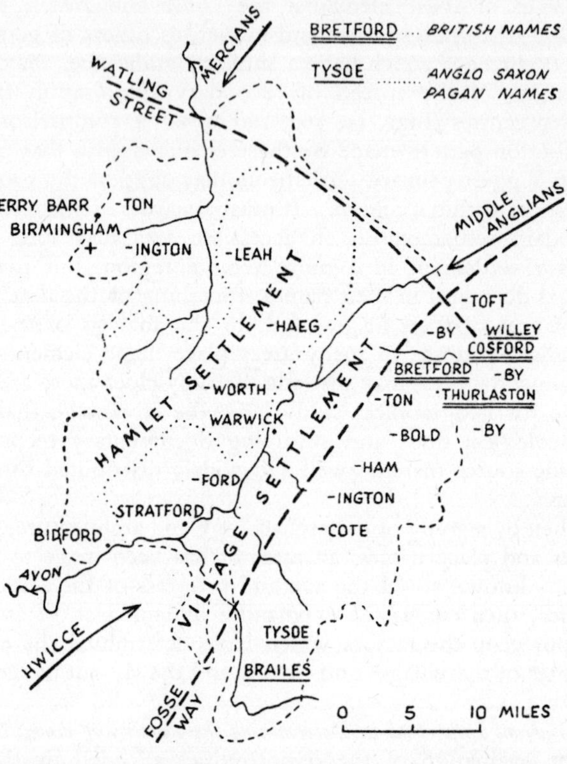

Fig. 17. The progress of settlement in Warwickshire. The place names by their distribution and form suggest the colonisation of the area by three groups : the Hwicce and the Middle Anglians (-bold *is a Middle Anglian element*) *in the Avon valley, and the Mercians in the Blyth valley. The settlement of the Avon valley with surviving British names, Anglo-Saxon pagan names and more numerous* -ham *names would seem perhaps to be earlier than the settlement of the Blyth with its more numerous* -tons. *The area of hamlet settlement with names with clearing elements shows the colonisation of the Forest of Arden. Scandinavian elements* (-by *and* -toft) *are found only near Watling Street.*

strong relief, abundant surface water, and thin soils except where well-drained deposits of fine sand or silt occur in old lake basins or in river valleys. Lowland Britain has a cover of newer rocks not yet wholly stripped off the old floor ; the edges of the more resistant layers of limestone, chalk and sandstone stand out as long lines of low hills, locally steep, overlooking vales worn in the clays. In Lowland Britain conditions of soil and water supply emphasise the division into hill and plain, for soils are light and surface water scanty on the pervious sands and limestones, and soils are heavy and surface water abundant on the impervious clays.

Dry Foundations

In Highland Britain the outcrops of resistant rock provided sound stable dry foundations wherever man might wish to build. Building stone was abundant too ; each village, even each farm, has its quarry obvious nearby. In Lowland Britain the uplands of limestone, chalk and sandstone provided dry sites and good building stone too, but in the clay lands, in the river valleys and above all in the marshes the search for dry foundations may have been all-compelling in choosing a place to live. Here too building materials had to be made from mud and clay, reeds, rushes and timber. These considerations in part explain the even scatter of the farms and villages in upland areas and the way in which in the vales of Lowland Britain the villages are often irregularly spaced, strung out, for example, along the valley on surviving patches of river terraces above the alluvial floor, or perched on hills or spurs above the plain.

Water Supplies

On the old resistant rocks of Highland Britain easily accessible water is also abundant (Fig. 18. 1). Perennial running surface streams and numerous springs associated with the faults in the complicated structures that result from a long, eventful geological history give man a wide choice in selecting his dwelling places. This is another fact that allows the scattered settlement of the Highland zone.

In the lowland zone the problem of water supply is greater. On the dry ground of limestone, chalk and sandstone, surface and near surface water may be scanty. The rain sinks in and the water may be stored too far below ground for men without

① WATER SUPPLY, HIGHLAND ZONE: RIVER AND FAULT SPRINGS.

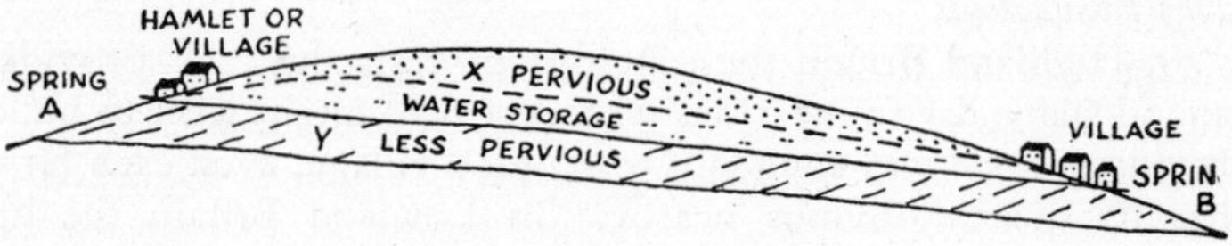

② SPRINGS AT OUTCROP OF LESS PERVIOUS UNDERLYING ROCK.

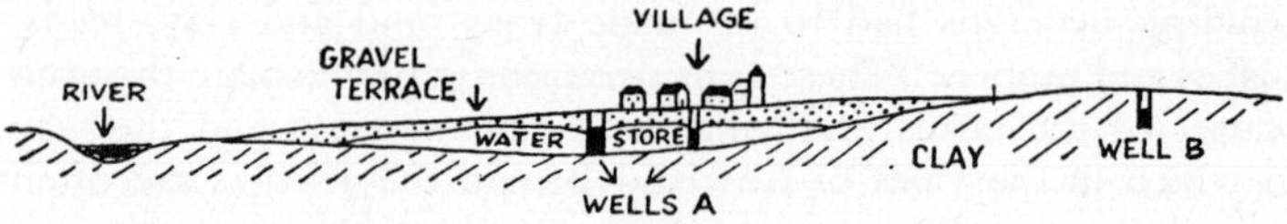

③ WELLS IN A PATCH OF RIVER GRAVEL ON IMPERVIOUS CLAY.

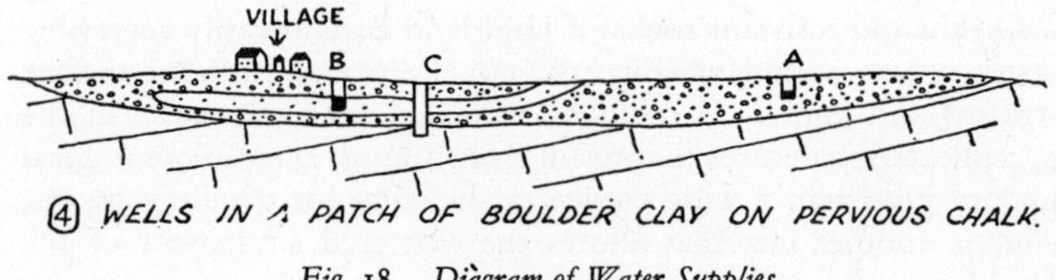

④ WELLS IN A PATCH OF BOULDER CLAY ON PERVIOUS CHALK.

Fig. 18. Diagram of Water Supplies.

elaborate boring and pumping equipment to reach it. On the wet ground of waterlogged clays, there may be plenty of surface water, but not much in one place and what there is, muddy and unattractive. Surface rivers in chalk and limestone country are few, in the clay vales they are often sluggish ; springs and shallow wells are therefore the important source of drinking water. Springs are thrown out along the outcrop of the junction between two bands of rock through which water seeps at different rates (Fig. 18. 2). Water does not seep out constantly all along the outcrop but issues at particular points where for one reason or another (Nature is almost infinitely variable) there is least resistance to its flow. These points were often selected by man for his villages ; in many parts of southern England village after village will be found along a spring line. Fig. 14 illustrates this ; the reader can probably call to mind many other examples. Shallow wells sunk in patches of sand or gravel may provide water both in areas where impervious clay outcrops generally or where dry chalk or limestone forms most of the surface. Gravel patches lying on impervious clays may be full of water at their base. Rain sinks into them and is held up by the clay below. A well (Fig. 18, 3) sunk into the gravel at A will reach water at no great depth and as water is drawn from the well more water will seep into the shaft laterally from all parts of the gravel patch to replace that drawn off. If a well is sunk into the clay itself at B the water may seep into the well shaft through the small pore spaces of the close packed grains so slowly that the well cannot provide a constant supply. A gravel patch lying upon rock through which water sinks rapidly will also often provide water in a shallow well (Fig. 18. 4). These gravel patches, whatever their origin—some may be remnants of a river terrace, others a scrap of an old beach, others a patch of boulder clay left by a vanished ice sheet—are rarely uniform in texture throughout, and thus rarely allow water to pass through at a uniform rate, and at the same rate as through the pervious strata below In Fig. 18. 4 water would pass most rapidly through the chalk, more slowly but still freely through the lenticular patch of sand, and more slowly still through the clay. A well sunk in the boulder clay at A would hold water, though if the clay were fine in grain with but little gravel intermixed water would seep in slowly. A well sunk into the sandy lens at B would have a better supply ; the water percolating in at its surface would be held up by the clay below but would move freely laterally and thus the well at B would draw upon the whole lens for supply. If a well were deepened in an effort to gain more water until it was sunk as at C through sand and clay to the chalk below, it might become dry as the water trickling in through the sand and seeping in through the clay might flow away faster still through cracks (joints) in the chalk at the bottom of the well. On heavy clay lands, and on limestone, chalk and sands, villages are often sited on patches of gravel since water can there be obtained from shallow wells. The supply of water in these wells is however limited, and the villages dependent upon them are not as a rule so large as those that have grown up near a powerful spring. These well villages are not only usually smaller than the spring line villages but also, in contrast to the linear arrangement of the latter, they appear like the villages of the resistant and impervious rocks of the Highlands to be scattered at random. They are scattered, but not at random.

The physical conditions here discussed governing building sites and water supplies have nothing like the overriding importance today that they once had. Now buildings can be rafted on shifting foundations and water can be led in pipes from distant storage lakes or pumped to the surface from bores sunk into water-bearing strata deep underground. Nevertheless, settlement patterns of the present day often show a detailed relationship to the minutiae of local geology which determines dry and stable foundations and the occurrence of surface and near surface water, a correspondence which is surprising if present conditions alone are considered but is readily understood when it is remembered that the pattern may be one inherited from the period of colonisation.

Soil Conditions and Farming Practices

The settlers chose dry sites on which to build not only with one eye on a good water supply, but also with the other seeking suitable soils on which to raise flocks and herds and crops. Ideas of what physical conditions were good and what not so good to provide water changed little if at all during the period of first settlement, but a great change would seem to have taken place in the appreciation of soil conditions. It is probable that the majority of the migrants who arrived in Britain in the Neolithic and Bronze Ages were primarily herdsmen though they cultivated a little grain ; the Iron Age groups were primarily arable farmers and the Anglo-Saxons at any rate were equipped with a more efficient plough than their predecessors. These facts strongly influenced the choice of places to settle.

The pre-Roman settlers sought dry pastures that would not poach in the rainy climate of north-west Europe, and soils easily worked with hoes and digging sticks and light ploughs. Cultivation with hoes and light ploughs broke up the surface sod and helped evaporation and percolation by stirring up the surface soil, but did little to assist the run-off of surface water. It was most successful therefore where slope encouraged surface run-off or where the underlying soil would let water pass freely through it. In Highland Britain sheltered valley sites open to sun and yet protected from strong winds and free from frost pockets were most sought after ; there pasture was most abundant and the best land could be cultivated to grow the limited crops. Here where slope helped run-off, waterlogging, except in local hollows where peat bogs formed, did not trouble the farmer in spite of the high rainfall of the west. In Lowland Britain, Neolithic and Bronze Age finds are most abundant on light soil uplands. Salisbury Plain, the Dorset Downs, the Cotswold Hills, Northampton Heights, Lincoln Edge and the Yorkshire Moors and the Berkshire Downs, the Chilterns, the Lincoln and Yorkshire Wolds show a relatively high density of finds in this period (Figs. 3 & 5). These areas were probably preferred not because they were naturally free from trees, but because once the trees were cut down their soils which were well drained suited best the practices of primitive farmers. How many of our modern villages in Highland Britain are on sites already established in pre-Roman times it is impossible to say, but it is clear that in general the villages and farms of Lowland Britain show in their distribution no preference for the very light soils, in fact the intermediate and heavier soils are evidently more favoured.

It was the introduction of a new plough that changed the

pattern of settlement in Lowland Britain. This plough was fitted not only with a share to make a groove, as had the light plough introduced from the Mediterranean by one or other of the Celtic groups, but also with a coulter (a knife) to cut the sod and a mould board to turn it. It seems probable that the

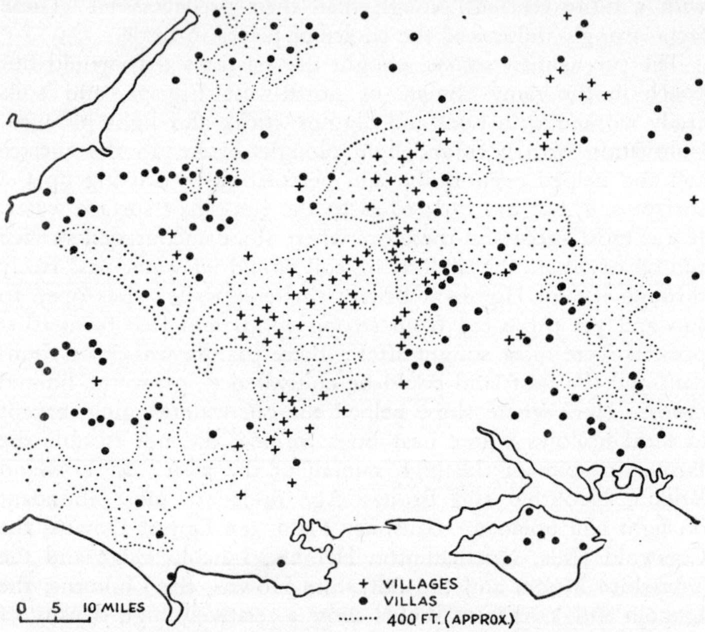

Fig. 19. Distribution of Village and Villa Settlement of the South Western Chalklands in Roman Britain. The villages which are probably pre-Roman settlements are found on the whole on higher ground than the villas. The Roman villas show perhaps the beginning of the colonisation of the heavier soils associated with the use of the mould board plough.

mould board plough in its essential form was first used by the Teutonic tribes of Western Germany. Whether it was an invention made independently from a digging tool, or whether it arose through modification to meet local conditions of the light Mediterranean plough or of a primitive plough in use among the Slavs, is uncertain. When the mould board plough was first introduced into England is again a matter for conjecture. The Belgae, who if not Teutons had relations with Teutons, and who crossed to Britain in the century before the Roman invasion, may have used it. The Romans must have known it, and the large fields so often associated with Roman villas in this country may have been laid out with it (Fig. 19). The Anglo-Saxon farmers if they did not introduce it certainly spread its use wide over the land. As we have seen the great majority of English villages bear Anglo-Saxon names; it is to the choice of Anglo-Saxon farmers that so many of us owe our place of residence to-day, so it behoves us to understand their choice. Their

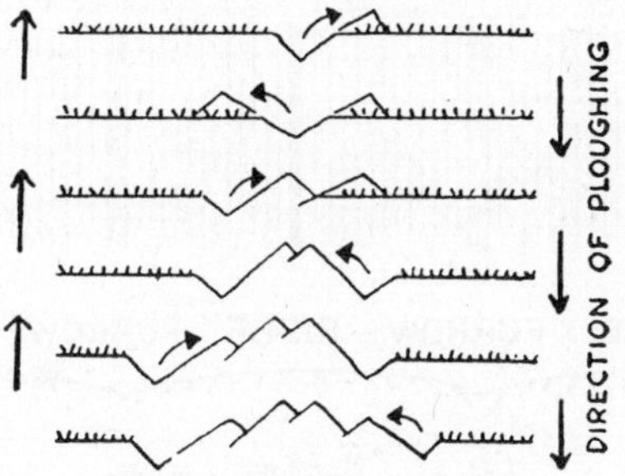

Fig. 20. Opening up a Top.

success as colonisers, the permanence of their village sites, and the good heart in which their farming practices maintained the land for centuries is at least in part the result of their adoption and use of the mould board plough. This process of ploughing lies at the root of so many features of the present landscape, and has so largely determined the modern settlement pattern, that it is therefore worth while to consider it in some detail.

The coulter fitted in front of the share cuts the mat of grass stems and roots and allows the easier entry of the share. The mould board fitted behind the share turns over the sod, thus exposing a greater surface area to evaporation and drying. If the overturned sod is not to lie on an unploughed strip of earth the line of the first furrow must be ploughed twice, once in each direction, and then the plough driven round it to throw two sods to the centre to form a "top." Ploughing is then round and round this top to form a ridge (Fig. 20).

When the ridge is wide enough a new "top" is opened up parallel to the first and half the width of the ridge away. The plough is driven round and round this top until the second ridge meets the first one. Along the junction of the two ridges the plough will have worked twice, the first time turning the sod towards the first ridge, the second time turning it to the second ridge thus making a double furrow. A ploughed field is thus thrown into a series of ridges :—

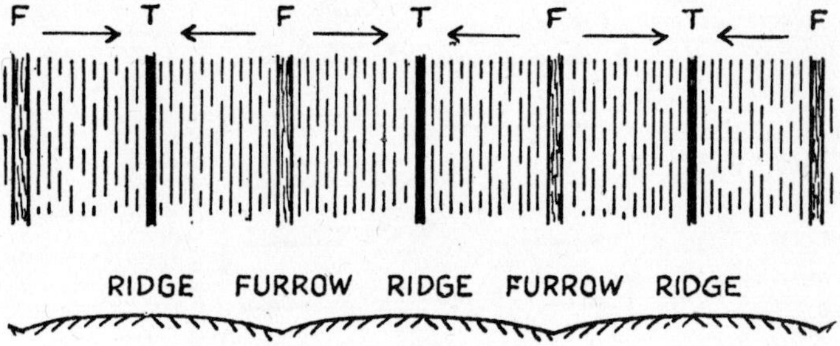

Fig. 21. *Ridge and Furrow.*

The mould board plough made of wood shod with iron was heavy to drag, especially if the slope was marked or the soil sticky. Eight oxen traditionally formed the plough team though the number varied. The plough and the plough team was an unwieldy unit; once it got going it was wasted effort to stop and start again unnecessarily, so that, where space allowed, the length of the furrow was that along which on a given slope with a certain stickiness of soil the team could pull the plough without stopping to rest. The "rest pause" was used to turn the plough, moving it round the end of the ridge with the share out of the ground ready to plough along the other side. When the width of the ridge became so great that the plough was driven without working further than the need for rest demanded, then a new ridge was set out. The traditional furrow-long (furlong) became 220 yards, the traditional width of the ridge 22 yds.;

thus evolved the common land measure of an acre 4,840 sq. yds., the area of the usual unit of ploughing, the strip. In practice it is found that furlongs and acres had different local values. The length of the strip depended upon the slope of the ground and the heaviness of the soil; the steeper the slope, the heavier the soil, the sooner the beasts needed a rest, the shorter the furlong. The width of the strips varied with the need for drainage. Before the nineteenth century when special field drains were made, the double furrows between the strips formed the only channels to carry off the surface water. Ploughing was usually carried out up and down the hill so that the furrows would be most effective. The heavier the soil, the flatter the land, the greater the danger of waterlogging, the narrower were the ridges made. They could be made not only narrow but high. Fields were ploughed more than once, sometimes three and four times in the season that they were fallow. The normal practice at the first, and perhaps the second, ploughing was to set out the new tops on the site of the old furrows, restoring the original lay-out in the final ploughing before sowing. The units of land holding were the strips so their position had to be a fixed one. This kept the land fairly flat. If however at every ploughing, or at the majority of ploughings, the new ridge tops were always set out along the old tops the soil became piled high on the ridges and gave what nineteenth-century farmers called "high-backed lands". Most people can call to mind seeing, before the war at least, ploughlands long in grass showing these steep ridges. These lands have perhaps not been ploughed since the introduction of field drains below the surface made high-backed lands unnecessary, and the use of seed drills, harrows and reapers made them troublesome.

Farmers equipped with a mould board plough no longer obtained their best yields in Britain from the drier soils. They sought the heavier loams and the clay lands in preference to the lighter soils, for these soils, with sufficient soil water to prevent summer drought and to allow continuous chemical change to replace used reserves of available minerals for plant food, could now be kept free from waterlogging with efficient ploughing. They gave good yields even with frequent cropping and allowed a relatively dense permanent settlement of agriculturalists. The farmers sought the plains and valleys too because they wished to lay out fields as large as the lie of the land and the resources of the community would allow, since the cumbersomeness of the plough with the large team of oxen meant that steep slopes and small areas were difficult to work. In these lowlands because the size of the fields and of the plough team made common labour profitable, they settled in villages, not in isolated independent farms and hamlets, and in villages as large as water supply would allow. Villages were established near a strong spring, or where several wells could be sunk, just any little well would not do. In these villages the manor, the church, the farms and the cottages were all part of the village scene. The most accessible and fertile land around the village was laid out by the plough in strips arranged in two or three large fields, and these fields were cultivated in rotation. The strips as ploughed were allotted one by one to lord and parson and every villager, each his share, so that each man held an equal number of scattered strips in each of the common fields. Arable land was obviously not all the village needed; in fact the extent of the arable in any one village was ultimately limited not as a rule by the extent of suitable soil for cultivation but by the proportion of the land of the community that must be left in grass to feed the beasts needed to pull the plough. The beasts grazed on the stubble after harvest and on the fallow arable before ploughing but this was not sufficient for them. The wetter parts, land near a stream for example, were used as meadow and mown for hay; downs, heaths and woods provided most of the rough grazing. The technical "waste"—heath, marsh and wood,—was valued not only for grazing but also for the provision of food and raw materials. The marsh provided fish and wild fowl; the heath and wood birds and other game for food. The marsh, heath and wood provided turf and peat and brushwood for fuel, reeds and rushes and brushwood for thatching and fencing; these materials mixed with clay often were the building materials too. Timber was used for the framework and floors of many houses, and the roofs of most buildings. Wood was needed for most farm implements and furniture. In the days when roads were few and transport difficult and costly, few daily necessities except salt and iron were traded widely ; each settlement tried to be as near self-sufficing as possible. Thus if arable could not be indefinitely extended at the expense of pasture, the pasture could not be extended beyond a certain point at the expense of the waste. Isolated fields and pastures were often reclaimed from the wild, woods hunted, streams fished, salt evaporated, and iron worked some distance from the village. It must have been convenient sometimes in these outlying places to build a barn to store grain, a shed to shelter beasts, perhaps a hut or two occupied at first in busy seasons, until from such small beginnings in some places a hamlet permanently inhabited grew. In frontier districts where there was room for expansion several such hamlets might form a group, assert rights over a certain area, build a new church, establish their economic and social independence, and thus a new village rather different in form from the parent would be born.

Not only the position of the villages in Lowland Britain but the pattern made by parish boundaries can now be appreciated. The variation in size is in part a reflection of accessibility, in part of fertility. The first to colonise often established themselves thickly on the " best " bits first reached, and these favoured spots were divided up among them in small parishes while later settlers scattered more widely in the wilderness, which though potentially just as fertile was for the moment less wanted. Early big parishes of this kind were frequently divided later. The parishes of north-west Essex are perhaps a good example (Fig. 10). Settlers coming in by the Stour and the Colne settled thickly once above the coastal marshes in the valleys of these rivers and their tributaries. Here the parishes are small. In the valleys of the Upper Stort and Upper Cam settlers coming in from the Thames valley by the Lea and the Stort and from the Wash by the Ouse and the Cam again settle thickly and the parishes are small. But in between these areas the parishes are much bigger even today, and their names suggest that they once were bigger still : pairs of names, Great and Little, are very common. The region formed by these large parishes, is covered with boulder clay and therefore the soil though locally variable in quality is on the whole good. Perhaps being rather heavy, at least in parts, it was not valued by the early settlers quite so much as the loamier soils of the valleys ; but it is more probable that the position off the main routes on the divide between the main lines of approach supplies the real reason for the later and sparser settlement that the large parishes suggest.

The shapes as well as the sizes of the parishes are illuminating. Where there is little variation in soil the parishes tend to be approximately round or square more or less symmetrically disposed about the village. Most of the parishes of High Norfolk and Suffolk are like this. But where the quality of the land is variable, the parishes boundaries tend to be arranged so that each villages has a share of the different soils. Where the structure is a simple tilt or fold and different rocks outcrop in linear bands, the parishes are often a series of long narrow rectangles cutting across the outcrops. The parishes of Lincoln Edge show this arrangement, two great groups meet along the crest. Each parish of the eastern group runs from crest to fenland, including within its boundaries a stretch of downland for pasture, of arable along the dip slope and of fen for meadow in the valley of the Witham ; each parish of the western group lies half upon the downland, half in the Trent valley. The villages of both series lie on the spring line halfway along the long axis of the parishes. The parishes of the chalklands of eastern Cambridgeshire show a similar arrangement (Fig. 11.B). In narrow river valleys the parishes tend to be arranged transversely across the valleys ; in the upper and narrower parts running right across the valley, in the lower and wider part in a double series each occupying one side of the valley, the river itself forming a common boundary along one of the short sides of the parish rectangle (Fig. 12). This arrangement too gives an economic variety of land ; upland pastures and water meadows with well-drained gentle slopes between for arable. In areas where one patch of soil is particularly desirable, the parish boundaries often make peculiar patterns so that each may have a share. On the Cambridgeshire chalk five villages share an outlying patch of Fenland particularly valued no doubt for its hay meadows and summer pastures as well as its reeds and rushes, turf, fish and water fowl (Fig. 11.A). Where villagers could not satisfy their needs in the immediate area, they sometimes acquired rights to pasture beasts, cut wood, evaporate salt or mine iron, for example, in an outlying area. Sometimes several villages each some distance away parcelled out among them an area that possessed a valuable local resource. The existence until very lately of small areas detached from the rest of the parish mark these old rights of common, in nearby, but not adjacent, areas.

The enclosure of the land, the breaking up of the open fields and common pastures of the village lands into individual farms, when it came, brought with it a certain dispersion of settlement. New farms were built conveniently placed near the centre of the new farm blocks of land. Enclosure was a slow and to a large extent a piecemeal and continuous process, but in much of Midland England at any rate it did not take place on any large scale until the late eighteenth and nineteenth centuries. Enclosure modified but did not alter radically the pattern of rural settlement.

The coming of the Anglo-Saxon farmers with their heavy plough and their preference for flat expanses of fertile loam and clay soils changed the pattern of Highland settlement much less fundamentally. In north and western Britain some of the coastal plains and the lower river valleys were well suited to cultivation with the mould board plough; sufficiently flat to allow the lay-out of large fields, sufficiently fertile to make relatively intensive farming worth while and to support big villages. Some of the bigger villages of the coastal lowlands of Devon and Cornwall, for example, may owe their origin to colonisation by Anglo-Saxon farmers or to the spread of their practices to British groups who adopted Anglo-Saxon ways. Recent research has found signs both in documents and on the ground of the lay-out at one time in these areas of arable land in two or three big fields. However, in some places in the west, even where physical conditions suggest that the use of the mould board plough would have been possible and profitable, big villages are rarely found. Here perhaps there was later and less intense colonisation, perhaps at a period and in a manner that did not favour co-aration; or perhaps the persistence of an earlier tradition firmly established accounts for the characteristic scene. In yet other areas slopes were too steep and land too infertile to lay out big fields and to sustain a regular rotation of two years' cropping with only a one-year fallow to rest the land. Here the best parts were ploughed, perhaps with a light plough, perhaps with a caschrom, that peculiar cross between a plough and a digging stick, an implement still in use in the North-West Highlands of Scotland and in Western Ireland in the nineteenth

Fig. 22. The Caschrom.

century. Professor Orwin suggests that the caschrom may have been the response in the Highlands to the introduction of the plough in the Lowlands. The instrument (Fig. 22) was pushed in with the foot and flicked over to the right by a turn of the wrist and thus a shallow furrow was driven across the field. The work was always done down hill and the field " ploughed " from right to left. The field in the west, however ploughed, was often continuously cropped until the yield fell very low, when it was allowed to revert to grass or heather for a long period and a new field cleared to take its place. Sometimes one field more fertile than the rest or near the settlement, the "infield", was kept always in cultivation by manuring it with all the waste of the village. In the Highlands, however, the growing of crops probably always took second place to the raising of beasts; the acreage of arable both in "the infield" and in the temporary brakes which together made "the outfield" was small in comparison with the acreage of meadow and pasture and rough hill grazing. Settlements remained small; a hamlet by the infield, and then to cowsheds and sheep pens on distant pastures were added cottages for cattlemen and shepherds and isolated farms grew up with small irregular cultivated fields near the farm house and wide stretches of moorland around. The parcelling out of land into individual farms has often followed closely upon or accompanied colonisation, and therefore in these areas enclosure brought little change in the pattern of settlement (Figs. 9 & 10). In these poorer regions a bigger area was needed to support the community than in more fertile areas and pressure on the land was less. Parishes were therefore large. If a map showing the parish boundaries of, for example, Cornwall, Cardigan, Cumberland or Northumberland is examined a striking difference is seen between the size of the parishes of the coastal lowlands and of the hill country inland.

These conditions, pastoral farming with infield-outfield cultivation with its hamlet settlement and large parishes, though characteristic of Highland Britain is not confined to it. In the poorer soil regions of the Lowlands, where these were cultivated, it is found too: in parts of Yorkshire, in the Sherwood Forest, and in the East Anglian Breckland for example. In fact Breckland is named after the practice of making temporary "brakes" for cultivation.

The great variety of villages and parishes is then no chance occurrence, but is the outcome of the way in which man has worked over many centuries in harmony with the great variety of nature. In former days village life was at one and the same time more uniform in response to a simpler economy dependent on the local supply or the local demand, and more variable because more sensitive to changes in physical conditions which contemporary techniques could do less to modify, than is village life today. But village life today nevertheless springs from and is built around that of the past.

Knowledge of English farms and villages helps in the understanding of European ones, and to see the European countryside throws a new light on the geography of rural England, for many features of the " English " scene are but European ones transplanted. Across the wide sweep of the North European plain from the Loire to the Elbe, village sites, village plans and village lands were chosen and designed only a little earlier than, and in much the same conditions as, those in England. Here in the lowlands, houses and farms are grouped together in villages with, before the enclosures, the arable land arranged in two or three large common fields around them. Within this area the Low Countries form an exceptional region. In the Low Countries as the result of extensive draining and reclamation in the Middle Ages the villages and their lands have a different look; isolated farms are perched on islands of drier land, and villages, with their houses strung out along the banks of a dyke and their land stretching from the dyke across the drained to the unreclaimed fen, are common. In the uplands, in the Ardennes and the Vosges, in Brittany and in the Central Massif there are as in Wales and Scotland and Ireland smaller hamlets, more and older isolated farms, and evidence of a long pastoral tradition.

But in southern and eastern Europe settlement patterns contrast with Britain. In the Mediterranean country people live both more widely scattered and more herded together. Little hamlets are frequent; but even more striking are the big villages, more like little towns to northern eyes than true villages, often perched upon a height, perhaps the more easily to guard against foes, whether the rapidly rising river flood, the " marsh miasma ", or the Phoenician, Greek, Saracen, or Barbary pirate. The summer drought is long, the autumn rainfall heavy; the slopes are steep, the soils thin and light. Frequent cultivation with a matchet, frequent stirring with a light plough keeps the soil broken up, favours percolation in wet weather and helps ground water to rise in dry. Both implements could be used efficiently in small fields, and fields often had to be small where slopes were steep and rocky. The variety of Mediterranean crops make a number of small fields of more use than two or three large ones; vineyards, not barley fields, provide the staple drink; olive trees supplement the fat supply derived from milk and butter and cheese; orchards and gardens hold an important place in agricultural economy. The steeper upper slopes and the marshy lands are used for pasture; often the two kinds of pasture are used in tandem by transhumant flocks, summering on mountain alps, wintering on river plains. The small yield per acre and the necessity to use odd scraps of good land wherever found encouraged the building of scattered farms and hamlets; these in turn encouraged the growth of a few bigger settlements, each serving as the centre for many smaller ones. Again troubled times encouraged communities of sufficient size to defend themselves. Scarcity of water during the long dry summer limited big settlements to the few places where a reliable and copious spring flowed.

Eastwards across the Elbe, in the Oder, the Vistula and the Danube basins the nucleated villages of the lowlands and the hamlets and scattered farms of the upland are found too, but there are also new elements that are alien to English eyes. On land cultivated since prehistoric time and long since cleared of its forest cover, the Franks and Saxons colonising eastward found familiar and congenial conditions and laid out farms and villages as in their homeland. But just as the Celtic way of doing things, modified no doubt where conditions were most like the fertile parts of the lowlands of north-west Europe by contact with the Teutons, persisted in some places in the west, so in some places in the east the Slav way of living, here more, here less modified by Teutonic ways, persisted. Even where Teutons settled in numbers they sometimes took over as going concerns earlier villages; there would not necessarily be any need to build anew. In many areas beyond the Elbe, Teutonic settlers in the twelfth and thirteenth centuries cleared great forests and drained marshland and brought land under the plough for the first time. In many ways the colonisation of these areas resembles the settlement of new lands overseas rather than the gradual nibbling away of the remnants of the waste that went on in French and English lands. The land was given in huge grants to princes, dukes and prelates and a deliberate policy was pursued of trying to attract settlers. A proclamation in 1108 by the leading bishops of Saxony runs:

> "The Slavs are an abominable people but their land is very rich in flesh, honey, grain, birds and abounding in all products of the fertility of the earth when cultivated so that none can be compared unto it. So they say who know. Wherefore O Saxons, Franks, Lotharingians, men of Flanders most famous, here you can both save your souls and, if it please you, acquire the best land to live in."

Wherever there were disputes in the towns of Flanders, wherever harvests failed and the price of corn rose high, wherever rivers flooded and seas inundated the lands, there agents of the colonising lords were to be found extolling the attractions of "new" lands across the Elbe. Their activities remind us of the "newlanders" five hundred years later who, often in reality agents of shipping companies in need of passengers or of planters short of labour, posed as successful colonists and drew pictures of America that rivalled in attractiveness the Elysian fields themselves. Just as the oppressed and discontented and the war victims of eighteenth-century Europe responded to their propaganda and sailed westward across the Atlantic, so, as the chronicler tells "*cum equis et bubis cum aratris et plaustris*" the colonists of the twelfth century trekked eastward across the Elbe. The "colonial" villages were and are very different from the older ones. The houses, or farms rather, were built roughly along the line of the road, a line frequently determined by the run of the valley floor or the main drain. The village lands were not laid out in common fields and pastures, but to each dwelling land was allotted, usually running back in a strip into the wild, first the garden, then the arable, then the pasture and finally the wood lot or the fen, and more could be taken in as need arose. As new settlers came, they built their farmhouses and cleared back or drained a strip of land at one end or the other of the row.

In attempting to see the landscape as it is today rooted back in the age when its now familiar and accepted features were first planted, it is perhaps possible even if the landscape is old to teach ourselves, as we have been trying to do, to visualise something of the more obvious material conditions that must have exercised a wide and general influence on the planting and thus have affected the design as it developed and matured. But it is extremely difficult to imagine the emotional setting in which men in this distant scene chose places in which to settle down. It is impossible to *know* anything at all of the innumerable accidents, material and immaterial, that must have been the immediate cause that led settlers as individuals to set forth to seek new homes, guided their voyaging, and determined their actions on arrival. The best way to realise the importance of the accidentals, many of them trivial in themselves but with far-reaching geographical consequences, the best way to feel the atmosphere of the pioneering days is to turn to the study of colonising ventures of Europeans overseas in recent times. These ventures are fully documented and if our primary purpose is to increase our understanding of the geography of our own land then the journals and letters of early settlers in temperate lands, and perhaps, since most nearly approximating to the conditions of the age of the settlement in England, of Anglo-Saxons in North America, will give most help. In William Bradford's *History of the Plimmoth Plantation* and in John Winthrop's *Journal* can be read accounts of the early days of settlement in New England. It brings very near the picture of Anglo-Saxons settling the English lowland though, of course, the difference in time meant a difference in the knowledge and equipment of the settlers. But perhaps not as great a difference as might be thought; spades and hoes and primitive ploughs and harrows were all these early New Englanders had. Their farming practices were less advanced than those of their contemporaries in England; lack of equipment and the temptation to exploit the stored resources of virgin land encouraged primitive ways. It is recorded that in 1637, seven years after the arrival of the first colonists, there were still only thirty-seven ploughs in use in all the immense area controlled by the Massachusetts Bay Company.

The Writings of Captain John Smith, edited by E. Arber, illuminates the setting of the very different settlements of Virginia. The difficulties that faced the ill-assorted, ill-equipped mob of settlers in the early days of the colony are vividly described. The environment was unfamiliar and too many " adventurers " were determined to get rich quick by seeking " gold and rubies lying upon the shore " rather than to settle down to the hard work of making a living by cultivation of an alien soil. John Smith entreats his Company " when you send again . . . send but thirty carpenters, husbandmen, gardeners, fishermen, blacksmiths, and masons, and diggers up of tree roots, well provided than a thousand such as we have ". The introduction of tobacco planting put the colony on its feet, and, in spite of every effort on economic and moral grounds to encourage a better balanced agriculture, tobacco soon dominated the life of a wide region. Land was first taken along the rich valley bottoms because tobacco to grow well and yield foliage of good flavour needed deep rich soil, and since the whole crop was exported and many necessities, even a large amount of food, imported, valley bottoms within reach of navigable water were the first settled. Tobacco was a crop exhausting to the soil so that the policy of the colonists was, as one of them put it, to " weed out, wear out, and walk out ". Single planters thus became possessed of large areas of land, and the planters' farms with the huts of their indentured labourers and later of their Negro slaves isolated in the midst of their tobacco lands became the characteristic unit of settlement. Plantations dominated the scene, towns were few and villages as England and New England knew them never grew.

Further west, farms and townships replaced both villages and plantations. Two novels, H. C. Brown, *Grandmother Brown's One Hundred Years*, and L. Bromefield, *The Farm*, both to some extent biographical, help the imagination to grasp the process and the problems of settlement in the Ohio valley. W. P.

Webb in *The Great Plains* and E. Dick in *The Sod House Frontier* give vivid descriptions of the difficulties and dangers facing the colonists who established ranches and homesteads in the West. Many diaries tell of the excitements of the mining camps of the 'forty-niners in the Sacramento, of the 'fifty-niners making "Pike's Peak or bust", of the seekers after precious metals in other remote valleys of the Rockies.

The new settlements established reflect not only the new physical environment and the immediate preoccupations of the settler but also his cultural background, adjusted perhaps to very different conditions. The colonist has frequently a nostalgic desire to keep some of the traditional features of the houses and villages of the "old" country though they may fulfil no purpose now; often he has also a reformer's zeal to be bound no more by other habits that have long irked him though they might still have served him well. In so far as the raw materials of the new country, and some perhaps imported from abroad or from home, allow, in so far as his knowledge, skill and determination go, the colonist combining the traditional and the functional constructs as nearly as possible what he considers the ideal home to meet his needs in his new land. And his needs are as varied as his ideas and his environment.

Map Exercises

Suggestions of maps the reader may like to make of his area to illustrate the chapter.

1. Map of the parish boundaries to show their pattern.
2. Map of present settlements.
3. Map of settlements on earliest edition O.S.
4. Map showing the towns, villages and hamlets from Saxton or Speed.
5. Map showing the vills named in the Domesday Survey. (In the north the earliest documentary mention of place names.)

6. Map showing the dates of foundation of parish churches.
7. Distribution map of place names containing pagan elements, and of *-ingas* names.
8. Distribution map of place name elements suggesting primary settlement, e.g. -ham, -ton, -borough, -bridge names.
9. Distribution map of place name elements suggesting secondary colonisation, e.g. -stead, -stow, stoc, wic, cote, etc.

CHAPTER V

THE SETTLEMENT OF THE LAND: TOWNS

ONLY rural settlements, farms, hamlets and villages have received attention so far, but in a country like Britain where, according to the census of 1951, for better or worse some 80% of the population of England and Wales now live in urban communities and 70% of the population of Scotland in Cities and Boroughs, towns interest many of us. Towns are not just overgrown and glorified villages. A town is not to be defined by size alone; it is essentially different in function from a village, though as Maitland reminds us the student of medieval towns and boroughs has often "fields and pastures on his hands". The definition of a town is notoriously difficult. It may be defined in the legal sense by its borough rights or municipal status, in the economic sense by its market facilities, or in the social sense by its cultural amenities. A town in the geographical sense serves in some capacity—defence, administration, production or trade—a wider area than the village; it is the centre of many villages and in return for the services it renders it, to a great extent, depends directly or indirectly on the countryside for food. The circumstances that give rise to towns, and thus many of the factors that must be considered in the study of their situation, site and plan differ from those already considered for villages. Towns have an even greater individuality than villages and it is even harder to make useful generalisations. What can be said with certainty is that to study the geography of a town effectively it must be examined in relation to the conditions of the period of its origin and growth; the study is essentially one in historical geography.

British towns have a long history; how long is a debatable point. Roman Britain had real towns: for example fortresses like Caerleon and Carlisle, administrative centres like Camulodunum (Colchester) and Verulamium (St. Albans), ports like

Regnum (Chichester) and Dubris (Dover), and the spa town of Bath. But, with the decay of Roman life in the northern provinces of the Empire, the Roman towns declined in prosperity and whether town life continued in any of them through the disturbed period of the Anglo-Saxon invasions is not known. It has been argued that in London, in Lincoln and in York, for example, continuous occupation takes us back to Roman times. Most of the Roman town sites are certainly sites of modern towns, but this may reflect the enduring attraction of the situations the Romans chose, and perhaps the temporary attraction of the Roman sites as quarries providing later builders with ready-cut stone, rather than the occupation of Roman towns as live communities. The lack of correspondence between the plan of Roman streets and surviving streets, with a few exceptions—e.g. Chichester and Chester, it is suggested, owe their rectangular plan to the survival of the Roman design—argues for the revival of town life on the Roman site rather than for its survival there from Roman times. Professor Stephenson, the American pupil of the Belgian authority on town origins, Professor Pirenne, argues by analogy with Professor Pirenne's view of conditions in North-West Europe, that English town life did not revive until after the Norman Conquest. He sees English towns with very few exceptions emerging as trading settlements established outside the walls of Norman castles. Documentary and archaeological evidence however would suggest that, whatever may have been the conditions during the most disturbed period of the Anglo-Saxon conquest, town life had developed in various forms long before the arrival of the Normans. Some towns are of course of recent origin; most industrial towns, many residential centres and holiday resorts have as their nucleus an old village perhaps, but none of the characteristics of a town until the late eighteenth, the nineteenth or even the twentieth century. Some few towns like New Winchelsea in the thirteenth century, the garden cities of Letchworth and Welwyn in the twentieth century and the "new" towns of this present decade are examples of planned communities, laid out to a design to meet the needs of urban living from the beginning.

Medieval towns were by present standards small; most towns

have grown greatly in the late nineteenth and twentieth centuries and it is salutory to be made fully aware of the very recent date of the present pattern and scale of urban living. Some few towns remain in medieval form and serve to point the contrast. Rye and Winchelsea, once busy Cinque ports but left high and dry by a receding sea, are fossil towns, their function gone. Rye (Fig. 23) with its Market Place, High Street, Mint, Strand and Quays has the lay-out and character

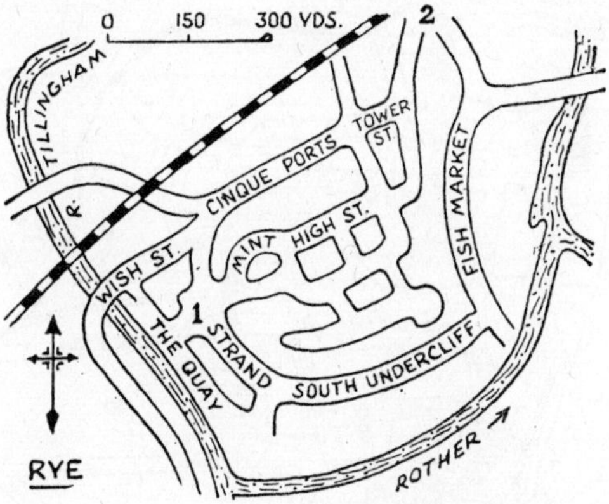

Fig. 23. *Town Plan of Rye.*
1. *Site of Strandgate.*
2. *Site of Landgate.*

of a town; for all its size it is no village, and its one-time purpose is clear. King's Lynn (Fig. 24) with King Street and Queen Street running parallel to the river front from the Tuesday to the Saturday Market Place, lined by imposing merchant houses with their warehouses on the Quays, with its South Gate and traces of its Wall to landward, retains the characteristics of a medieval port, though a busy local market town has here continued to serve a prosperous countryside and the built-up area has increased greatly in the modern period. Lavenham, Suffolk (Fig. 25), is another town arrested in development; like Rye and Winchelsea it is a survival with no real modern

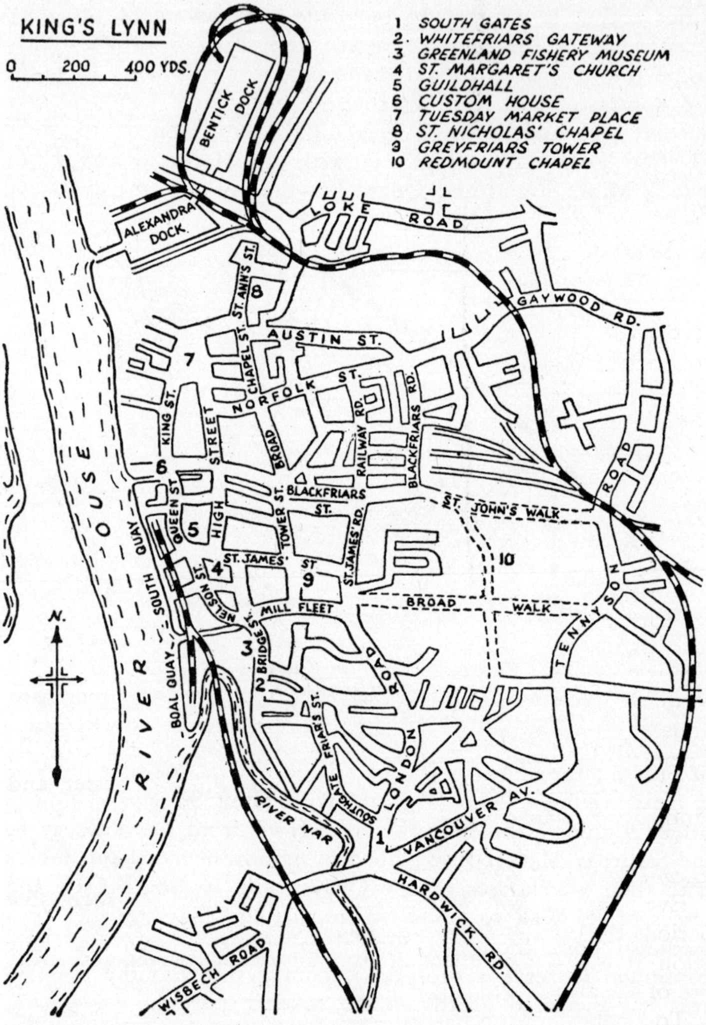

Fig. 24. *Town Plan of King's Lynn.*

function. It is a rare example of a medieval industrial town, one of the centres of the Suffolk cloth manufacture. It stands today almost untouched, a miniature town on two little spurs in the valley of a headstream of the Brett. It is a dual settlement, the church and rectory are on the summit of the southern spur with the Hall on its flank, the traditional grouping of an agricultural village; but there is also a commercial and industrial quarter, the Market Place crowns the northern spur with High Street, Prentice Street, Shilling Street and Water

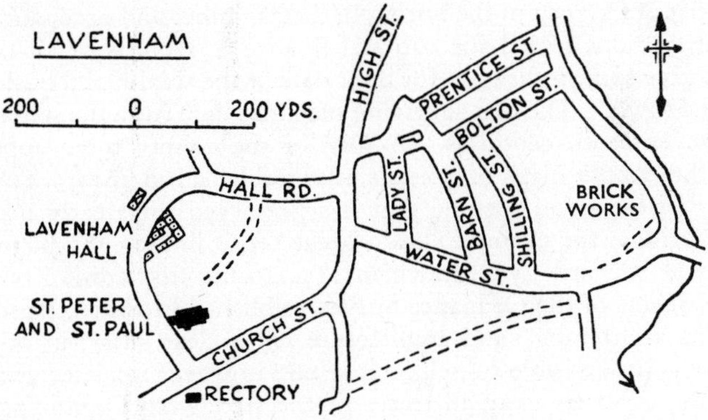

Fig. 25. Town Plan of Lavenham.

Street around it still with their medieval cottages, merchant houses, Guildhall and Wool Hall to emphasise their character. There could hardly be a greater contrast between these towns and, for example, Liverpool and Glasgow, Manchester and Huddersfield, their counterparts today.

Local Sources

Every reader will gain much by examining for successive periods the plan of a town he knows well in relation to its situation, its site and its place in the political, economic and social life of the area.

To establish the town plan for different periods is the first task; the sources for doing so differ greatly from town to town.

For some, documents and maps both published and unpublished are numerous, for others few. The Domesday Survey is rarely very helpful. The information given of Domesday boroughs is difficult of interpretation, it is uncertain whether all, or if not all what proportion of, the houses and lands of a borough were recorded, and there are but few clues to their distribution. However, it is worth looking at the record, if it exists, to see what it does say. For the medieval period the grant of market rights, of minting rights, of burghal or civic status; the tolls collected to pave streets, build and repair bridges and walls; the list of freemen of the borough, their number and occupations; contributions levied for national finances; records of all these will give some indication for their date of the wealth and business of the town. The first surviving map may date from the sixteenth or seventeenth centuries. A plan of the county town appears on the county maps of Speed's atlas published in 1611. Towns vary in the good fortune that has preserved their early maps; Cambridge, for example, has a great series linking the pictorial plan of Richard Lyne drawn in 1574 to the first edition of the 6 in. sheets of the Ordnance Survey published in 1885. In some towns archaeology and architecture tell a clear story, in others there is little visible to help. It is rare however for later generations to sweep away all traces of the streets and buildings of their predecessors. For one thing, to do so is often very costly; it is easier to build anew on fresh fields than to clear old sites. For another, pride in and affection for an architectural inheritance is strong and, even if it hampers modern planning, an effort is made to incorporate within the new the valued features of the old. Some indication of the extent of the town of Roman date, if one existed, may sometimes be traced on the ground. The Romans were solid builders. The limits of the medieval town, its walls and gateways are often still impressive monuments, and even when their masonry has vanished their position can often be traced by examining the run of the streets and the surviving street names. The medieval walls and gates of York and Lincoln, Chester and Berwick-on-Tweed for example are still impressive; the walls and gates of Chichester, Shrewsbury and Winchester can be followed in the streets. Short and narrow streets of complex pattern sometimes stand out in contrast to wider and more symmetrically arranged thoroughfares in a plan of streets and show up startlingly clearly the old core of the town. The distribution of churches of a medieval date indicates too the extent of the contemporary town; and their number size and magnificence, and the small size of the parishes which supported them, often most forcibly suggests the concentration of wealth in the settlement. York and Norwich are famous not only for their cathedrals but for the number and beauty of their parish churches; in cathedral towns perhaps this does not surprise us, but the number of great medieval churches within a small area in Cambridge, Stamford and Bristol for instance calls sharply to mind the strength of local commerce then. The Georgian squares, the Regency terraces, the Victorian villas, the semi-detached residences, the council houses and the pre-fabs carry on the story of town growth through the centuries.

Using these and other sources a series of maps should be made illustrating the extent and plan of the town at significant dates. The earliest area of settlement can perhaps best be appreciated by mapping the recorded sites of the archaeological exhibits in the local museum. The position and scatter of these finds may suggest what it was in the situation and site that first attracted settlers. A map of surviving medieval buildings should be constructed and compared with the earliest extant town plan. The town as it was before the first period of rapid expansion early in the nineteenth century if a suitable contemporary map can be found, or if not as it was on the first edition of the plan made by the Ordnance Survey, may be the next in the series. For many towns it is valuable to map the extent in the decade before the 1914-18 war and then the extent in the early 1930's to link the nineteenth-century plan to the present one. To appreciate the physical geography of the site it is a great help to make on a scale suited to the size of the town—a 6-in. or even a 25-in. scale for small towns, a 2½-in. or 1-in. scale for large ones—a relief map and, if possible, a subsoil map, omitting all but a few key streets and buildings needed to give orientation and position to the mind's eye. In a built-up area the detail of streets and buildings obscures the relief on a map, and the uneven sky line of roofs distracts the eye from the ground, so that the essential relief of a town is often not well known, unless slopes are considerable. It is, however, paradoxically, often on low-lying sites where changes in relief are very slight that these changes have most effect on the building plan. It may be that contours as given on the usual Ordnance Survey sheets are not sufficient to show the details of relief that have affected the town plan. Spot heights taken from the 25-in. scale plans or, by the courtesy of the borough engineer, from his maps, will allow the interpolation of form lines at closer vertical intervals. The local 6 in. geological sheets, supplemented when possible by observation wherever and whenever excavations on building sites are open, will give the data for the construction of a map of the subsoils. The general situation of the town can only be appreciated in relation to the communications which link it to the area in which it lies. Some suggestions for the mapping of local communications for different periods are made in Chapter VIII.

Factors Affecting Town Growth

The considerations of sound foundations and good water supplies which affect the choice of village sites apply with even greater force to towns. Town dwellers congregate in larger numbers, and a situation that attracts them is by no means always blessed with a desirable site. A situation that attracted because it was easily defended, or because it was a focus point for traffic where alone the sea could be reached or a river easily forded or bridged, might by that very fact prove to be a site restricted by steep slopes or marshy ground. Lincoln and Durham perched in a commanding way high above their valleys, Norwich and Shrewsbury in their river meanders, Bedford and Bridgwater at their fords had but limited areas easily built upon. The hill town of Old Sarum with its two Cathedrals and its Norman Castle was gradually abandoned after a new Cathedral was built in the thirteenth century upon the plain of Salisbury. The effect on the town plan of good and difficult building sites, before modern engineering did so much to obliterate the differences, is often clearly reflected in the relative dates of the different quarters of the town. Earlier built-up areas often coincide with gravelly and sandy subsoils, the later ones with clay. Anomalous isolated blocks of later houses or even of open ground within an area early built-up will often be found to mark out a patch of sticky clay or perhaps of water-logged alluvium or peat.

If dry and level ground for building was limited in many towns and cities, then as population grew houses became more and more crowded on the restricted area, and water supplies became not only short but contaminated. The river itself, shallow wells in river gravels, or local springs in pervious strata, which were the usual sources of early town supplies, were easily polluted by a large group with no proper methods of disposing of waste and sewage. The descriptions so frequently found of the filthy conditions of the streets of medieval towns and the measures so often, and therefore presumably so ineffectually, prescribed by local authorities or even by King or Parliament to bring remedy, afford abundant evidence of how contamination occurred. Undoubtedly a certain immunity to water-borne diseases developed among those who survived in town populations, but the death rate in towns was before the late nineteenth century significantly high.

The great increase in the population and area of many towns from the middle of the nineteenth century frequently reflects four changes: the enclosure of town fields or the acquisition of estates hitherto in private hands which released for building sites desirable land hitherto used for agriculture; the use of difficult sites by expensive draining, by "making" of ground or by building on piles; the engineering of modern water supplies and sanitation systems; and the coming of the railways. There is cause and effect here. Increasing pressure of population made expensive building schemes worth while, and in turn the more numerous the houses gathered together the more imperative was the need for good organisation and careful control to provide a pure water supply. The cholera epidemics of the 1830's and 40's roused public opinion about the danger of an insanitary water supply, and many towns established corporation water works about the middle of the century. The greater the amount of building land, the better the water supply, the greater the increase of population with a falling death rate and increased immigration, the greater the demand for food. A big town could not be entirely supplied from the immediate neighbourhood; towns could not grow beyond a certain size unless on navigable water as London, Paris or Venice until the railways were built to carry in abundant and cheap food supplies and also to distribute the products and services which the town offered to pay for food. The railways too increased the area from which the town attracted immigrants, thus contributing directly to the gain in the rate of increase of its population.

The desirability of building sites, the availability of water supply and good communications did not however primarily determine the origin and growth of the town, though growth must take place within the limits imposed by them. The form of the town, like the form of the village, must be examined in relation to the function of the settlement. A town rarely fulfils throughout its existence one function only. No other geographical study illustrates more forcibly the strength of historical momentum. A town established to perform one task, takes upon itself often by virtue of its very existence other tasks in addition; and, if the need for one function passes, another comes to dominate the life of the place even if the site and situation is not entirely suited to the new role and there must be considerable adaptation in the design of the town. Only in a very few cases however does fulfilment of a later task entirely obliterate all the elements of the design evolved to meet an earlier one, and in studying the town, past as well as present functions must be considered. A frontier fortress becomes a market and a cathedral town as Chester and Carlisle, perhaps an industrial city in addition as York; a local market becomes a university as Cambridge, and also a great industrial town as Oxford; a port becomes a centre of industry too as Liverpool, Bristol and Glasgow have done; some towns are primarily industrial and commercial centres as Birmingham, Bradford, Leeds and Manchester for example; and capital cities add royal residences and seats of Government to all the rest; Edinburgh is a capital in miniature, London on a grand scale.

It is impossible here to do more than illustrate in the barest outline the close relationship between situation, size, function and structure by reference to a few examples: York, which has all the elements the historian demands for his concept of a town; Cambridge, typical of a local market but a special town too; the port of Bristol; the conurbation of Birmingham; and, as London is far too complex, the capital city of Edinburgh.

York (Fig. 26), situated in the midst of its fertile vale flanking the most important route to the North, has been built upon since prehistoric times. It lies at the confluence of the Ouse and the Foss where these rivers cut through a long ridge of gravels which curving across the low-lying ground drew most of the traffic crossing the plain. The Ouse, until the river flow was regulated in the eighteenth century, was tidal to the confluence and navigable to the Humber. The site then had obvious strategic and commercial importance. York was a British settlement, and when the Romans conquered the Brigantes, the most powerful tribe of the northern region, York (Eburacum) became the principal fortress of Roman Britain, the base of the legions which manned the northern wall. After the withdrawal of the Roman armies it retained, or soon regained, its powerful position. It was the capital of Northumbria at the height of its power; and when the Danes overran the north, that part of Northumbria south of the Tees became the Danish kingdom of York with York the capital. The area, remote and individual, retained a certain independence as a vassal state when the West Saxon kings shired the Midlands, and it was the centre of strong resistance to the Normans. It was captured by the Normans; William the Conqueror built a castle at the river confluence which was soon stormed and taken by the men of the North. In revenge the Normans virtually destroyed the town, but, rebuilt and refortified, it continued to be a military centre of great importance. But York was no mere fortress, it was early the centre of the northern church. In the fourth century a bishop of York, together with a bishop of London, is recorded

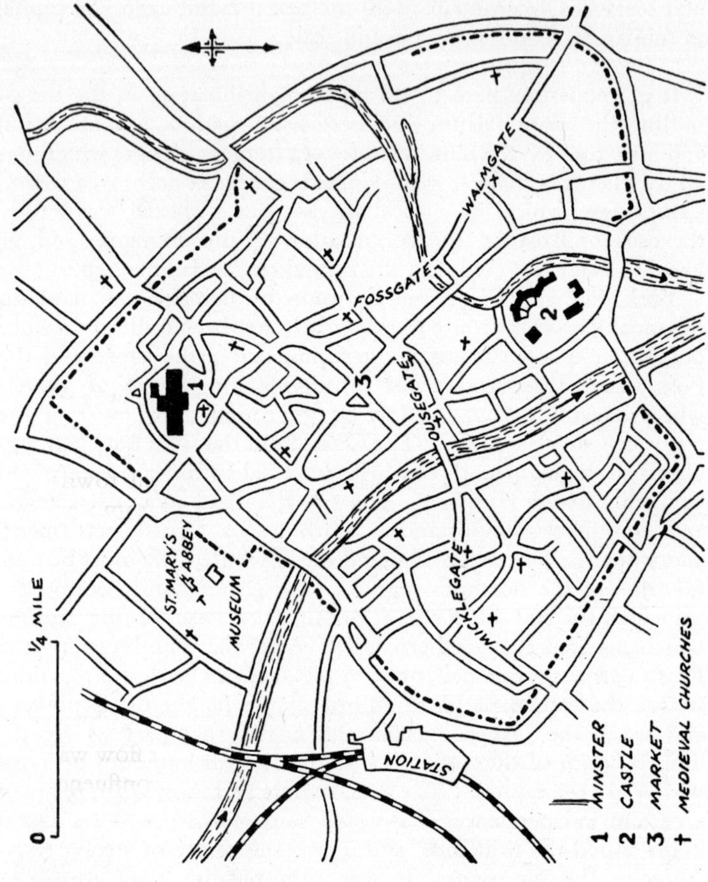

Fig 26. Town Plan of York.

as attending the Council of Arles. On the conversion of the Anglo-Saxons, King Edwin was baptised here in 627, and soon afterwards York was made the capital of the northern one of the two archbishoprics into which Britain was divided. Edwin began the building of a stone church on the site of the little wooden one in which he was baptised. His church was burnt down in

741 and replaced by the Saxon Cathedral, largely destroyed in its turn by the Normans ; a little walling of eighth-century date in the crypt of the present Minster alone survives. The present cathedral rose slowly, the choir is of the twelfth century, the

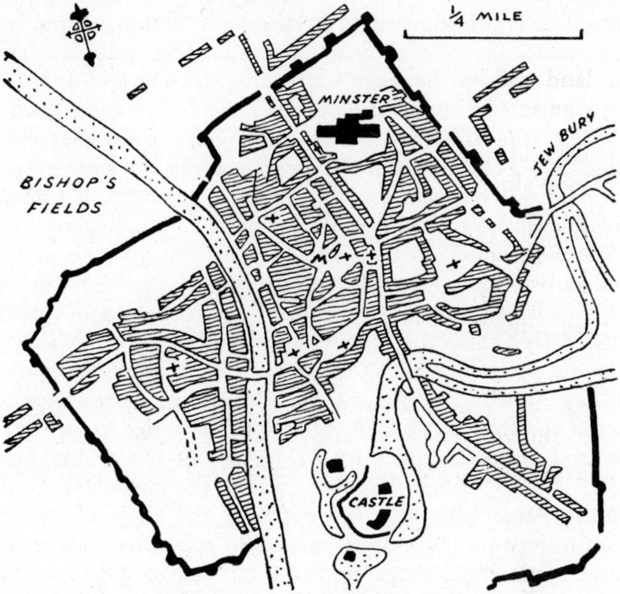

Fig. 27. *Plan of City of York, by John Rocque.* (*This plan and those of Figs.* 31 *and* 34 *have been drawn from* '*A Collection of Plans of the Principal Cities of Great Britain and Ireland; with Maps of the Coast of the said Kingdoms; Drawn from the most Accurate Surveys; In particular those taken by the late Mr. J. Rocque, Topographer to His Majesty*'. 1764.)

transepts of the thirteenth, and the nave was completed in 1340. The church and the cathedral on its site were built within the walls of the Roman town. From early times the farming wealth of the countryside and the traffic on the Ouse supplied a busy market serving the soldiers of the castle and the clergy of the Minster and of St. Mary's Abbey ; and by the twelfth century York had a cloth industry of international reputation.

The centre of the present town preserves these three elements: the Roman and Anglo-Saxon fortifications have all but disappeared but the remains of the Norman Castle and the thirteenth-century Clifford's Tower stands on an impressive mound at the river confluence in the south; the magnificent Minster, with its medieval glass still glowing in its windows, dominates the northern quarter; the Market Square, significantly called Parliament Street, lies mid-way between Castle and Minster, and little streets —the "gates" to commemorate the Danes—radiating from each link the three together. The town early grew beyond the circumscribed site between the rivers; along the Foss especially much land was too marshy to build on. Ousegate led down to the bridge across the Ouse and continued as Micklegate through a western suburb; Fossgate led to the Foss bridge and continued as Walmgate through the eastern quarter. The medieval walls, much of their fourteenth-century masonry still standing, enclose within them the Minster and its gardens, the Market, the Castle and the two riparian quarters. St. Mary's Abbey and its lands lay outside, within the shadow of the north-west wall. Between the fourteenth and eighteenth centuries expansion was slight and John Rocque's plan (Fig. 27) of the mid-eighteenth century shows York largely within its medieval bounds. Modern streets with new bridges over the river now encircle the old town running closely parallel to the walls; these streets mark abruptly the passage from the old city to the new. York has grown immensely since the coming of the railways re-emphasised its old function as a route centre. There are large railway workshops to the north of the city. York lost in the late medieval period its local pre-eminence in the cloth industry, but printing and bookbinding, glass making, chemical industries and the manufacture of cocoa and chocolate now provide work. The modern suburbs, housing the workers, have been built along and between the main roads linking York to Hull and the Humber, to the Vale of Pickering and to the towns of the Vale of York and the Eastern Pennines.

Cambridge (Fig. 28) preserves equally clearly its distinctive quarters illustrating the functions it has successively and simultaneously fulfilled. It lies where fen and upland meet, being

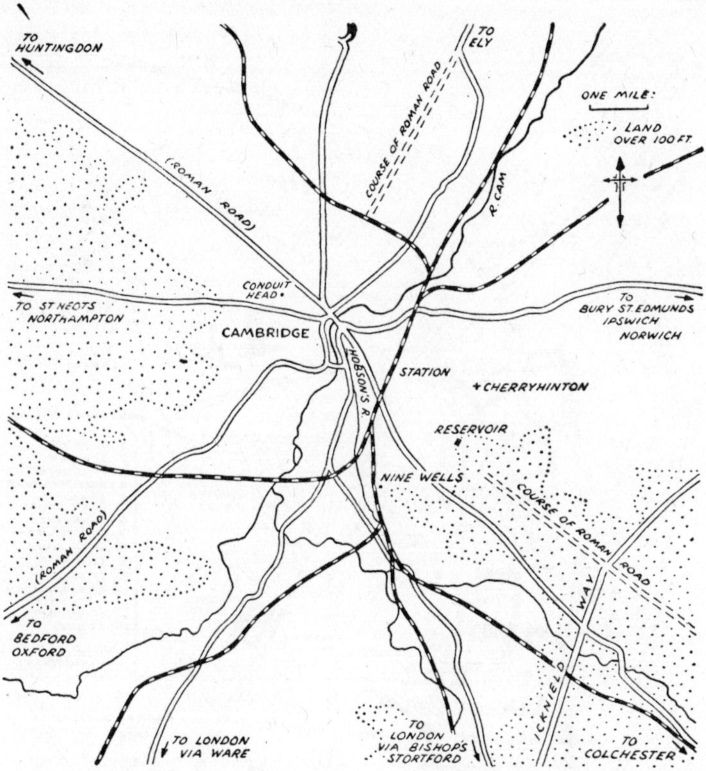

Fig. 28. The Position of Cambridge showing routes and water supplies.

the most northerly point where traffic from East Anglia gathering on to the Icknield Way could pass easily to the Midlands and northwards along Ermine Street (Fig. 50). A spur of chalk jutting north-westwards from the Chiltern scarp carried a dry route to the river bend where fording and later bridging, was possible. Outliers of chalk helped routes north-westwards from this crossing, and a route, made into a paved road by the Romans, picked its way along spreads of river gravel to the island of Ely. This crossing and these routes still dominate the town plan (Fig. 29).

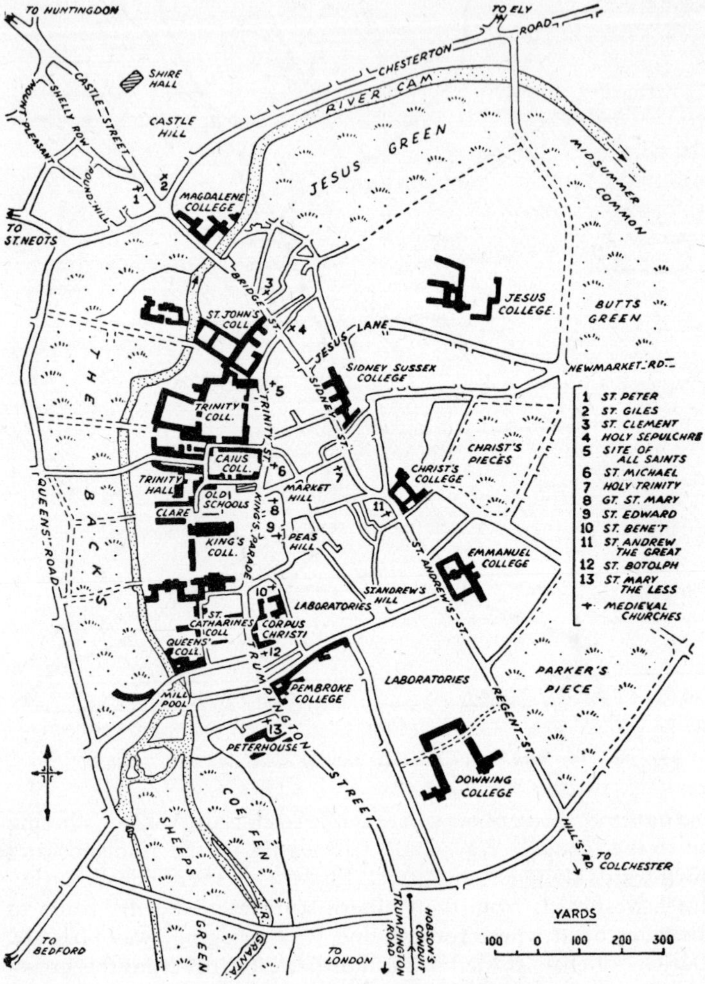

Fig. 29. Town Plan of Cambridge.

To the south of the river the heart of the town lies between the road from London and the Thames valley (Trumpington Road-King's Parade), the road from Colchester and Suffolk (Hills Road-St Andrew's Street), and the road from Norwich and Norfolk (Newmarket Road-Jesus Lane); the three roads gather into Bridge Street and the centre one is known to modern town planners as "the spine". The river was navigable to the bridging point and just beyond; boats from all the Fenland waterways, and small coasting ships too, tied up along the quays. The traffic of these land and water routes early brought traders to this site where fenland products could be exchanged for those of arable land; doubtless boat builders, cart makers, harness makers, smiths, cobblers and tailors not to mention inn-keepers were soon permanently resident to serve the travellers. Finds of prehistoric date are numerous around the crossing, but whether these are the surviving remnants of goods and chattels lost or jettisoned by travellers at a difficult point on a journey or the waste of a permanent settlement it is hard to say. The crossing, inhabited or not, was of sufficient importance for the Romans to build extensive fortifications on the flanks of the chalk outlier on the northern bank and Anglo-Saxon cemeteries of the pagan period near it suggest settlement around it then. Prosperity would appear not have have been unbroken, for Bede, telling of the monks of Ely voyaging upstream at the end of the seventh century in search of stone to make a coffin for their Abbess, describes Grantaceastr, where incidentally they find a Roman sarcophagus complete with lid, as a "desolate little city". Whatever its condition then, long before Norman times it revived. Irish merchants, probably Danes, are recorded trading here in the tenth century, and St. Clement's Church, standing on rising ground immediately to the south of the quays, to judge by its dedication to a favourite Danish saint may have been founded to minister to them. Two churches north of the river, St. Giles and St. Peter, have or had Saxon work in their fabric, and the Saxon tower of St. Benet's by the Market Hill stands still, an impressive monument to the early wealth of the southern quarter. One, perhaps both, of the churches by the outer gates, St. Andrew's by Barnwell Gate and St. Peter's by

Trumpington Gate (now St. Mary's the Less) may be pre-Conquest foundations. If so the settlement had reached approximately its medieval, even its eighteenth-century, area by the time the Domesday Commissioners arrived to count the 373 mansiones in the ten wards and to record that 27 had been destroyed to allow space for the construction of the Norman Castle. Medieval buildings clustered around the chalk outcrop of Castle Hill and on the " hills ", very low but for all that very important on this low-lying site, of terrace gravel in the meander, and along the hithes of the river front. The northern and southern part of the town may have once been separate. Bridge Street perhaps crept out towards the crossing over empty land; the haymarket lay near St. Clement's Church, and All Saints on the other side of the road was known as All Saints-in-the-Jewry or sometimes as All Saints-in-the-Suburbs. The haymarket, the beast market, and the Jewish quarters were frequently found on the edge of medieval towns.

What attracted the scholars? Did monkish stewards and itinerant friars come to buy at the market, stay to preach and teach, and in their turn attract other scholars? It is not known; but the religious orders took up residence in the town first, soon followed by secular groups. In the early days the scholars and their pupils lived as and where they could, then with generous endowments first hostels and later Colleges were built in which to live and work. The easiest building sites were occupied by the time the monks and scholars came. Two alternatives were thus open to them: to build on good gravel sites on the edge of the town as did the Dominicans on the site later used by Emmanuel College, and Pembroke and Peterhouse; or to squeeze in between castle and market along the edge of the river terrace, " making " ground and in part displacing wharves and hithes and streets and lanes leading down to them, to do so. The nuns of St. Radegund's (on the site later used by Jesus College), the Hospitallers of St. John (St. John's College), built on more or less empty and perhaps till then marshy land on either side of Bridge Street. Hostels and Colleges were built in and between the river lanes on the sites that were to become those of Trinity, Gonville and Caius, Trinity Hall and Clare. The Carmelites moved from a site beyond the marshes west of the river to a site on the bank of the Mill Pool by the small bridges, because in their old home floods so frequently prevented the monks reaching Cambridge to buy and sell, and their pupils reaching them for instruction. King Henry VI grandly swept away a parish church and many houses to acquire the land for his college, King's. Only three groups, the Franciscans on the site of Sidney Sussex, the Augustinians in Benet Street and the Guild of Corpus Christi managed to acquire land close to the heart of the town on Market Hill. Theirs was not land of the best; it was on the edge of a depression in the gravels marking a former course of the river, which was followed by an early defensive work known as the King's Ditch, a marshy and insalubrious streak that remained open ground until the nineteenth century.

Population grew in the seventeenth and eighteenth centuries but the area built upon and the character of the town changed little. Then in the early nineteenth century the town fields were enclosed, in the mid-nineteenth century the railway reached Cambridge, and in the late nineteenth century some industry of more than local importance was established. These changes transformed the town, but they added to rather than eclipsed the pattern already established. The enclosure of the town fields released to east and west large areas of gravel soils for building. Plentiful and cheap building land in the outskirts of the town not only relieved over-crowding—the rapidity with which new land was taken up was some measure of the need for that relief—but also prevented the destruction of the old town by continued overbuilding within it. In the middle of the nineteenth century in fact, some of the small houses which filled the Market Hill were accidentally burnt down and not replaced, and others, there and in groups which had been built in many yards and closes, were removed. Improvement of the water supply together with the increase in the built-up area allowed the growth of population. In medieval times the river and shallow wells in the gravels provided water; many houses and yards had each their own well and pumps tapping shallow wells were in public use around the Market Hill even in the nineteenth century. But some use was early made of springs outside the town. In the fourteenth century the Franciscan monks laid pipes which ran beneath the river to springs in gravels to the north of the town two and a half miles from their convent, springs near the present road, Conduit Head (Fig. 28). In the seventeenth century the channel now known as Hobson's River was artificially cut to lead water from springs in the chalk at Nine Wells to the south of the town, originally for the purpose of flushing out the King's Ditch. The water failed to do this effectively but Thomas Hobson, the famous carrier, fostered a project to pipe it to feed a drinking fountain in the Market Place. In 1853 the Town and University Waterworks Company was established and it brought an adequate and pure supply from copious springs in the water-bearing chalk at Cherryhinton. The coming of the railway might have destroyed the old town pattern; the railway authorities were anxious to run their lines into the very heart of the town and attempts were made to acquire part of the Backs or one of the Pieces on which to build a railway station. These plans however were resisted; the line was run along the height of river gravel to the east of the town and the station built upon the outskirts. This encouraged expansion to the east; commercial traffic decreased on the Cam as the railway took over the old function of the river as the highway for heavy and bulky goods. The coal wharves and grain hithes were forsaken for railway sidings, and the flour mills at work since Domesday times on the banks of the Mill Pool were replaced by factories near the station. The position of the railway aided the movement of population out of the old centre of the town; a footnote to the 1851 census gives specifically as the reason for the decline of the population in one river-side parish " many families have left as the Eastern Counties Railway is absorbing the trade of the Cam ". Land newly available for building and particularly sites near the railway attracted industry; industry for the most part linked to the area either by a local raw material, for example the cement works using a limey clay at the chalk marl/gault horizon, or by the special needs of the local market, for example the printing works and the scientific instrument companies. The industrial quarter grew then to the east of the town and has so far remained of moderate size. To the south and to the north and west residential quarters have grown rapidly, stretching out to, and incorporating within them several villages once distinct.

Thus Cambridge was a castle and a market, and a busy market too, long before it was a University town, let alone a railway junction and a centre of industry. The need for defence, at any rate by a castle on the river bank, has passed away; but Cambridge retains its other functions and as yet its plan. At the heart of the present town castle and market still dominate the scene. The castle with its ramparts as reconstructed by Edward I still determines the lay-out of the streets of the old town north of the river. Castle Street uses the old vallum to gain the easiest gradient up to the road to Huntingdon; on the west "Rows" of cottages and almshouses crown the ramparts the slopes of which, grass or slab covered, rise above the streets; on the east the Castle Mound looks down on the modern Shire Hall. To the south of the river the market place with the narrow streets and passages around it occupies the centre. People from a wide area still come to Cambridge, and come to it to buy and sell rather than to pass through it. The streets leading to the bridge are still crowded; the Market Hill with the shops and offices around it carries on business not only with the city but with the fen and upland farms and villages now as it always did, though the Cattle Market has moved from a site near the Market Hill to a site near the railway station. Men from all over Britain, from Dominions and Colonies and foreign countries, too, come to live and study in the Colleges which together form almost a closed ring around the market town. This old town is separated from the new not as in York by medieval walls and gateways, but by medieval pastures. Around the old town open to this day lie the "Backs" along the river alluvium, and the "Pieces" approximately coinciding with an outcrop of gault clay between the gravels of the meander core on which the old town was built and the gravels of a higher terrace on which was the abbey and village of Barnwell, now joined to Cambridge by nineteenth-century streets. If industry is restrained, if a recent tendency to give Cambridge yet another function as a centre of regional administration can be kept in check—and these are big "ifs"—then the growth in size of the town may be controlled, the traditional functions that it has been built to carry on fulfilled, and its surpassing beauty preserved and enhanced for future generations to enjoy.

The port of Bristol offers a great contrast to York and Cambridge. Both the latter are commercially and industrially of regional importance only, though each on account of one industry is a household word—chocolate and wireless come immediately to mind. Both have strong dual personalities; the one is as much a cathedral city, the other a university city, as a market town. Both are of limited size: York has a population[1] of 105,336, Cambridge of 81,463. Bristol with a population of 442,281 is far larger than either; and though it is the market of a fertile agricultural district, a cathedral city, a university town, and the centre of a famous industry, tobacco, it is essentially a seaport. It is worldwide trade that has determined its structure and character, and supports its great population. York and Cambridge ranked as seaports once, but the sea and sea trade never dominated life in them as it has always done in Bristol. How early were people living here is not known; the archaeologists have found evidence of prehistoric settlement not on Bristol's marshes but on the high ground above the Avon gorge at Clifton. The Romans too seem to have shunned the site; there are signs of a Roman camp, but one of no particular importance, at Sea Mills more than two miles down river from the heart of Bristol. The earliest certain evidence of Bristol's existence are two silver coins of the reign of Ethelred the Unready (979-1016) and others of the reign of Canute suggesting a pre-Conquest town already interested in trade. In the Middle Ages Bristol certainly served a wide hinterland; down river from Bath, along cart-tracks and bridle-paths from Cotswolds and Mendips came men to sell their cloth and corn and wool. Along the Severn came the barges loaded with the products of the West Midlands and the Welsh border, the wool and the corn, the butter and the cheese, the timber and coal, and again cloth. From across the Bristol Channel came iron and timber from the Forest of Dean, the hides and skins and cattle and once again cloth from

[1] 1951 Census. Preliminary Report.

South Wales, Somerset and Devon, the tin from Cornwall and fish from them all. Overseas trade in similar products was brisk with Ireland, and Bristol even knew Iceland's fish. The market for these goods lay partly in exchange between local group and local group, but a big share of the hides and skins, corn and wool, and particularly of the cloth collected in Bristol was marketed abroad especially in Gascony, Portugal and Spain. These countries had no important cloth industry of their own, therefore Bristol in the Middle Ages was never so important in the wool trade as were the east coast ports with their links with Flanders and Southampton with her Italian trade, but depended largely upon her cloth trade. Cloth from the fourteenth century to the sixteenth century dominated Bristol's export trade. Wine from Bordeaux and Bayonne, from Portugal and Spain was the great import, along with fruits, oil, salt, and many raw materials used in cloth-making, particularly " grain " and other dyestuffs, potash and alum as mordants. The local merchants who frequented Bristol must have returned home well satisfied with the great variety of necessities and luxuries they had been able to purchase in the city. The Cabots, contemporaries of Columbus, settled in Bristol in the fifteenth century when already Bristol men were discussing new lands and seeking for them in the west. Soon the Newfoundland Banks and the West Indian islands were frequented by her sailors and, though Bristol suffered some decline at the end of the Tudor period, with the opening up of the New World came new prosperity; Bristol's position and tradition encouraged her citizens to take a full share in the exploitation of the new opportunities for trade. In the seventeenth and eighteenth centuries, when Southampton had become " almost forsook and neglected " having lost her continental trade to London, Bristol was growing rapidly. Celia Fiennes visiting Bristol in 1696 says, " the town is a very great trading city as most in England, and is esteemed the largest next London; the river Avon, that is flowed up by the sea into Severn and so up the Avon into the town beares shippes and barges up to the key where I saw the harbour full of shipps carrying coales and all sorts of commodities to other parts ". In the eighteenth century the triangular trade across the Atlantic—

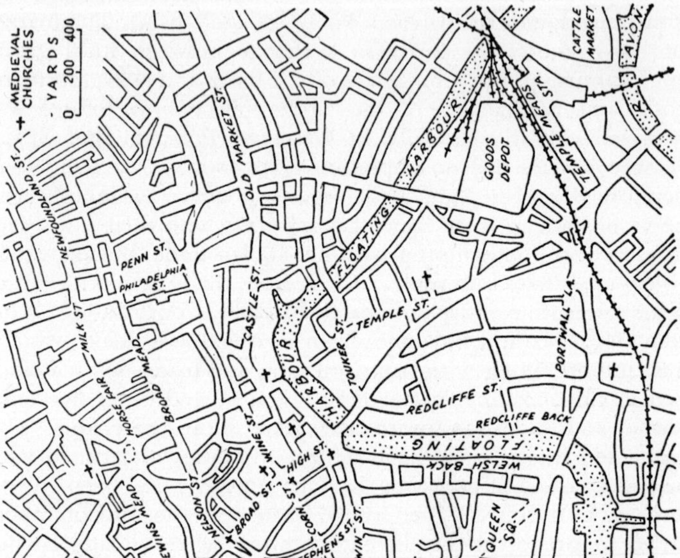

Fig. 30. Town Plan of Bristol.

the export of a wide variety of goods to West Africa, the transport of Negroes to the sugar plantations of the West Indies and to the tobacco colonies on the mainland, the carrying of sugar, molasses, and tobacco back to feed Bristol's processing industries—proved most profitable. In the early nineteenth century times were less good: the competition with Liverpool when it too entered the Atlantic trade, the abolition of the slave trade, and the decline of the West Indian trade called a temporary halt to Bristol's expansion. Liverpool, with a rapidly growing industrial hinterland manufacturing cotton, drew a greater and greater share of Atlantic commerce to herself. The opening of the canals linking the Staffordshire potteries and the Black Country iron industries to the Mersey before the cutting of the Trent-Severn links also abstracted to the newer port an important part of the trade of a region which had long been tributary to Bristol. Finally, Bristol had constructed in 1809 a new floating harbour and, to pay for it, she had to ask higher dock charges than were paid at Liverpool and Hull. By the end of the century, however, Bristol was again prosperous. The expansion of the Guinea trade had largely made up for the decline of the West Indian trade; railway building restored easy communication with the hinterland although the use of a 7-ft. gauge in the West somewhat restricted its scope; the construction of the modern docks at Avonmouth and Portishead in 1877-80 increased sea trade, a large share of the banana trade and of petroleum imports was added to her commerce, and new engineering industries were established in and near the city.

It is against this background that the geography of Bristol must be examined (Fig. 30). The situation offered great opportunities; the immediate site, on a deep waterway, sheltered from the open sea, on an easily defended promontory with a frontage on two rivers, offered advantages to a small group of boatmen and traders. The early settlement lies still clear at the core of the present town. Across the narrow tongue of land rising steeply between Frome and Avon ran four streets; the High Street and Broad Street and, at right angles to them, Corn Street and Wine Street (originally Winch Street), significant names. The circular curve of Nelson Street, St. Stephen's Street, and

Nicholas Street probably marks the course of the first town walls, maybe of Anglo-Saxon date. They did not long contain the town and a northern suburb was built and walled beyond it perhaps even before Norman times. The Castle, the only obvious sign of which is now in street names, was built by the Normans across the neck of the promontory on the east to guard the vulnerable land approach. Very soon this site was found too small; to the east beyond the Castle a new suburb grew with a market of its own, still commemorated in Old Market Street. To the north of the river Frome on the rising ground several monastic foundations were established, around which villages grew: the Black Friars (or Castle Friars) in Marshall Street just across the Frome, the Priory of St. James on the slopes of Kingsdown Hill, the Grey Friars in Lewin's Mead, the nunnery of St. Mary Magdalene at the foot of St. Michael's Hill, the Carmelites on the site of Colston Hall, and towards St. Brandon's Hill, St. Augustine's Abbey whose church was later to become the Cathedral. Monastic and abbey lands thus prevented the expansion of the commercial and industrial quarters of the town in this physically perhaps the most obvious, direction. There were two alternatives: to expand south of the Avon, and to extend the promontory site; both were taken. To the south of the Avon on the dry sandstone of the river meander core, a flourishing industrial quarter arose, how flourishing the great churches of St. Mary's Redcliffe and the Temple church (alas burnt out after an air raid in the 1939-45 war) show. The cloth makers gathered here along Temple Street and Touker Street, and in 1247-48 a stone bridge on which were shops and houses replaced an earlier wooden one to connect the suburb with the centre of the town. A wall was built across the landward side of the river bend; its line shown until recent redevelopment by Portwall Lane and Piper's Lane. In this new quarter Redcliff Back became one of the busiest of the quays. To extend the original site a new cut was made in 1239-40 along which the Frome was diverted across the marshland to the Avon. This greatly increased the length of the quays; those on the Frome came to be known as the Broad Quay as distinct from the Welsh Back, the Avon frontage of the reclaimed land. The area between the quays was marshy but on the drier northern end a new parish, St. Stephen's, was set out; Balanco Street (now Baldwin Street), along the outside of the old wall and following approximately the former course of the river, linked the Broad Quay to the Welsh Back, and Marsh Street and

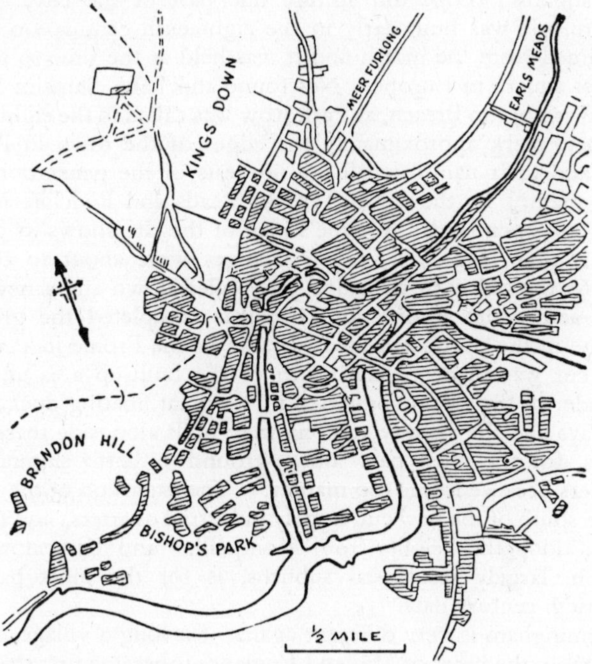

Fig. 31. A Plan of the City and Suburbs of Bristol, by John Rocque.

other little streets running down to the Marshland were built and protected on the south by a new wall. Broadmead, Horse Fair, Lewin's Mead and Milk Street preserve in their names the position of the meadows and pasture lands around the medieval town. Two extant plans, Hoffnagle's of 1581 and Millerd's of 1671, show relatively little increase or change in the area of the town. In 1735 Pope describes the city as " very unpleasant with no civilised company in it". "The streets," he says, "are as crowded as London but the best image I can give you of it is, 'tis as if Wapping and Southwark were ten times as big or all their people run into London." But in the eighteenth century, colonial trade well established, the town laid out new streets and squares; Pope did admire one part of the city, Queen Square. It was built early in the eighteenth century on a site reclaimed from the marsh and it was held at the time to be the largest square in Europe. Newfoundland Lane, Jamaica Street and Washington Breach, as Park Row was called in the eighteenth century, mark approximately the edges of the town in Pope's day, and their names recall the interests of the generation that built them. To the south Temple Meads and Redcliffe Meads were still open land. Rocque's plan of the city shows its extent at this time (Fig. 31). Great changes were about to come: in 1767 the old narrow bridge was pulled down and a new one built across the Avon. In 1809 was completed the greatest change in the port since the diversion of the Frome in 1239; a new cut was made to the south of the built-up area and the meander of the Avon converted into a great floating dock. The railways and the railway stations were built alongside these new docks, for the most part along ground hitherto shunned by builders because it was too marshy. The extension of the town to the south of the new cut, largely industrial quarters, and to the north along the heights from Montpellier and Kingsdown to Clifton, largely residential suburbs, is for the most part of twentieth-century date.

Birmingham is very different again; for long a village, small, remote, in the heart of Midland England, it became first the local market of its district and then, when technical change revolutionised the iron and steel industry and the communications of the area, it became a great industrial town, the centre of an immense industrial area. If commerce made Bristol, industry made Birmingham. Of Birmingham's earliest days nothing is known; its name containing the element *-ingas* suggests an early settlement but the first record of the village is in the Domesday Survey. It is then a tiny place with an assessment of four hides, land for six ploughs, a population of five villeins and four bordars, and a value of twenty shillings. The immediately adjoining manors of Aston, Selly Oak, and Erdington, now part of Birmingham, were at that time slightly bigger than it; but all the settlements of the Birmingham plateau were small in comparison with those of the Avon valley. By the fourteenth century, to judge by the 1327 Subsidy return, Birmingham had outstripped all the neighbouring villages and become a market town of comparable size to Droitwich, Bromsgrove and Warwick, though rather smaller than Lichfield and considerably smaller than Coventry. The explanation of the differential growth of settlements in this area lies more probably in variations in the use of opportunities made by man, than in variations in the opportunities offered by nature. The way across the plain of York, the way round the southern end of the Fenland, the deep and sheltered waterway leading inland from the Severn estuaries, with in each case an easy crossing place where marshes stretched for miles, these brought men again and again approximately to the sites of York, Cambridge and Bristol rather than to places nearby. But Birmingham, on the height of land where arose small tributaries flowing east and west, north and south to join only eventually one of the major rivers, had apparently nothing particularly inherent in situation or site to make it a place especially frequented. The rivers of the area were small and could be easily crossed wherever encountered; the land was rolling and wooded but slopes not so steep nor woods so thick that free movement was hindered; traffic was local, variable in route and unimportant in quantity and value. The main highways of prehistoric and Roman time which linked the populous south and west to the east Midlands, Lincoln and York, and the Thames valley to the Welsh frontier passed east and north of the Birmingham district (Fig. 17). The Fosseway ran east of the Avon using the rising ground of the limestone-sandstone scarpland, Watling Street ran from the Avon to the Trent valley by the cut through Cannock Chase; only Ryknield Street, perhaps marking a stage in the advance of the Roman frontier against the Celts from the Fosseway to the Welsh marches, crossed the area. It ran from the Fosseway at Bourton-on-the-Water to Doncaster, crossing the Avon at Bidford and the Trent near Alrewas, at both places spreads of river gravel narrowed the marshy valleys. Not even the Ryknield Way went through Birmingham; it crossed the Rea, Birmingham's river, at Lifford and the Tame at Perry Bar. By the fourteenth century this road pattern had changed if the Gough map may be relied upon; the main road east-west across the region had been deflected from the old line of Watling Street to pass through Coventry, and Ryknield Street now ran through Birmingham on its way from Worcester and Droitwich to Lichfield. According to Ranulph Higden Ryknield Street was one of the four great roads of England ". . . per Wygorniam, per Wycum, per Birmyncham, Lichfield, Derby, Chesterfield, Ekoracum usque ad ostium Tynae". Birmingham and Coventry then, off the early tracks, had become important enough in the fourteenth century to draw roads to them. Wherein lay the attraction of these places to men?

Birmingham lay upon a ridge of Keuper sandstone overlooking the little clay-floored valley of the Rea. It was a dry site yet with a copious water supply. A fault separating sandstone from marl threw out springs, and small bands of clay intercalated in the limestone itself held up perched water tables that fed shallow wells. In the valley below, the Rea was bridged using a little island in the stream; and by the bridge Deritend a little settlement grew up which, though on the far side of the river in Aston parish, was part of Birmingham manor. A mill was built downstream from the bridge, there were meadows along the river, cornfields on the valley slopes, and presumably plenty of wasteland in the yet thinly peopled area. The site was good undoubtedly but probably its physical characteristics and the opportunities they offered could be matched in nearly every neighbouring village. The policy or the good fortune of the lord, rather than innate natural qualities in situation and site, would seem to have fostered the early growth of Birmingham, and the success with which an initial lead once gained was maintained and increased is an excellent illustration of the geographical inertia created by the force of historical momentum. The Birmingham family was granted a market charter in 1166, earlier than any other village in the neighbourhood. A market charter by itself will not ensure the growth of a market town but

Birmingham, though probably even now still very small, was well placed to serve as an exchange centre for those villages of the plateau which lay too far from Droitwich, Coventry or Lichfield for any one of them to be convenient. It lay conveniently too between the richer agricultural lands of the south-east and the poorer pastures of the north-west. Records are not numerous but, to judge from those that exist concerning rents, grants, licences and court cases of the thirteenth and fourteenth centuries, it would seem that the lords encouraged or allowed their tenants to make good use of their privileges. The tenants early freed themselves from labour services, which allowed them to devote more time to making marketable goods. Plentiful waste allowed newcomers to obtain land for houses and closes. " The liberties and customs merchant of the market of Burmicham " became a recognised code, and by the fourteenth century grants of privileges are found no longer made to the " lord " but to " the Burgesses of the town of Bermingham " or to " the Bailiffes and Commonalty ". Cattle were brought to the fair in numbers ; woollen cloth, iron, steel and brass were sold in the market ; Birmingham had become a town.

The market doubtless encouraged local industry. By the time of Leland's visit in 1540 the industries which were to become dominant were well established. Leland's description tells vividly of the town and its life at this date and is worth quoting. He approached from Deritend :

> " I came through a pratty strete or evar I enteryd into Bremisham toune. This strete as I remember is caullyd Dyrtey, in it dwelle smithes and cutlers, and there is a brooke that devydithe this strete from Bremisham. Drytey is but a hamlet . . . and is clene separated from Bremischam paroche.
>
> " The beauty of Bremischam, a good market towne . . . is in one street goynge up alonge almoste from the lefte ripe of the broke up a mene hille by the length of a quarter of a mile. I saw but one paroche churche in the towne. There be many smithes in the towne that use to make knives and all manner of cuttynge tooles, and many lorimars that make byts, and a greate many naylors. So that a great parte of the towne

> is mayntayned by smithes. The smithes there have yren out of Staffordshire and Warwickeshire and sel coale out of Staffordshire."

Leland makes no mention of leather and cloth making, both old industries of the town; however, by the sixteenth century, smithing had clearly become the leading industry. Plentiful raw material, bar iron of good quality, could be obtained from the numerous iron works of the neighbourhood. By the end of the seventeenth century evidence suggests that the population of the one-time tiny village had risen to about 5,000. During the eighteenth century there was increasing concentration on "new" trades: the making of steel toys, of buttons and buckles and ornaments in brass, enamel or silver, and the manufacture of small arms gradually took precedence over smithing and the nail trade. The remote situation and the distance from navigable water tied industry to the production of small articles, light in weight depending largely on skill in production for their value.

Three things changed the possibilities of Birmingham's position in the late eighteenth and the nineteenth centuries. First, the coming of canals and then of railways gave her cheap and easy communication with the coalfields and the iron-working districts, and with London and the ports. Secondly, the invention of the steam engine, largely developed by Boulton and Watt at their Soho Manufactory on Houndsworth Heath north of the town, gave her a source of power which allowed mechanisation of industry far beyond what was possible using the tiny streams to turn water wheels. Thirdly, there came the engineering of water supplies on a large scale. Good though the local supplies had been, they were by the nineteenth century insufficient for the needs of the town. In 1826 a company was granted power to bring water from the Tame and in 1831 the Aston reservoir was made. In 1876 the water works were purchased by the Corporation and active steps were taken to improve and increase supplies still further. Public ownership enabled some 3,000 polluted wells within the city to be closed. By the end of the nineteenth century supplies from the reservoirs filled by the Tame and the Bourne were no longer sufficient; it was

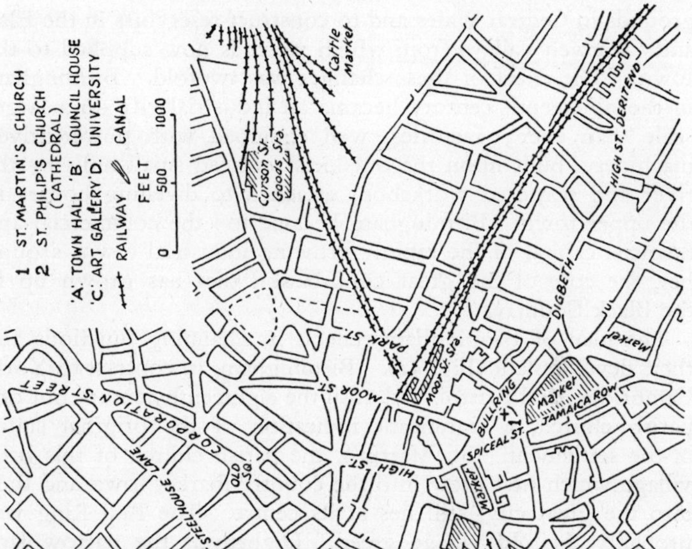

Fig. 32. Town Plan of Birmingham.

impossible to extend the capture area locally and as a result of increased building part of it was already suspect for purity. In 1892 an Act authorised the town to acquire extensive gathering grounds in Central Wales and to construct reservoirs in the Elan and Claerwen valleys from which water is now supplied to the town. The effect of these changes was twofold. Birmingham in the nineteenth century became an industrial city on a larger scale than ever; factories, well equipped with power-driven machinery, built upon the outskirts replaced small mills by the river and scattered workshops attached to dwelling houses in the upper town. Birmingham became too the commercial and financial capital of the rapidly growing industrial towns around her, the core of the great conurbation that has grown up in the Black Country.

The plan of Birmingham reflects its changing functions and their development (Fig. 32). Birmingham, in contrast to York, Cambridge, and Bristol, had until the eighteenth century but one parish church, a significant indication of the original status of the settlement. St. Martin's, the parish church of this little village which first grew into the country market town and then into the great city, still lies at its centre. The Bull Ring was probably the old village green. Digbeth is the narrow lane that led down to the river bridge and the group of houses at Deritend. Around the Bull Ring and along the High Street rising to higher ground, the markets and fairs were held and here the modern market halls were built in the nineteenth century. Jamaica Row and Spiceal Street suggest that the produce of distant lands as well as those of surrounding areas were bought and sold here. Here too were and are the shops and business houses, the centre of retail trade. Along the river were the tanyards and blade mills and smithies with workers' houses in "close courts" crowded thick around them. In the eighteenth century new building was carried on apace. In Digbeth and Deritend there was little change; here in the "watery parts" of the town liable to flood there was little room for expansion, but northward of St. Martin's church on the well-drained slopes of the Keuper Marl new streets and squares were laid out. Old Square justly famous, on the lands of the medieval Priory of St. Thomas, and Temple Row around the churchyard of St. Philip's, became the fashionable suburbs; here gathered the wealthier merchants and the growing group of professional men. The new church of St. Philip (now the pro-Cathedral) was consecrated in 1715; the Act of Parliament sanctioning the creation of a new parish reads, "the Towne of Birmingham being a Market Town of great trade and commerce was become so very populous, that, having but one church in it, it could not contain the greater part of the inhabitants . . . whereupon there should be a new church erected and a new churchyard set forth, and a new parish made". The new parish was rapidly built up; Park Street and Moor Street were largely built by 1730 and building was proceeding up Snow Hill, and the New Hall estate to the west of it was laid out. To the area east of Snow Hill the gun trade moved from Digbeth and Edgbaston Street and still remains here. To the area west of Snow Hill moved the prosperous toy, button and buckle makers and here to-day is still the quarter of the jewellers. The writers of the local directory published in 1777 states that:

> "the concourse of nobility, gentry, merchants and traders (as well foreigners as natives), who are continually flocking here is amazing, the former out of curiosity, the others in the way of business, who seldom return without expressing their astonishment and satisfaction at the number and ingenuity of the artificers, who may be said to be really in the possession of the philosophers' stone and literally to have the power of converting iron into gold without any considerable diminution of its weight."

However, a local history published in 1797, albeit largely a re-edition of an earlier work and therefore perhaps a little out of date, declares that "Birmingham is not a place a gentleman would choose to make his residence" and cites among its disadvantages "the close population, the noxious effusion of various metallic trades and above all the continuall smoke arising from the immense quantity of coals consumed". A strong movement of population to the outskirts has taken place in the nineteenth and twentieth centuries. At the end of the eighteenth century wealthy citizens were building houses a mile or more from the town presumably to escape from "the noxious effusions" and with the building of railways and the coming of the motor car residential areas and dormitory suburbs extend further and further. There has been, too, a move of industry to the periphery seeking space for large factories with if possible a canal or railway frontage. To the central industrial areas were added first compact areas around the town largely devoted to the manufacture of brass. Later factory areas stretched out in tongues along canals and railways, and were built around villages until then untouched by urbanisation. Here were housed the motor car and accessory industries. Growth along the periphery has been accompanied by changes in the central districts. The railways reached the town in 1837 and the first station was opened at Curzon Street. This proved too far out and in the 1850's the railways were brought into the centre of the town by tunnels, much slum property was cleared and New Street and Moor Street stations built. Further clearance of the central area has taken place; to improve the shopping centre many areas of squalid slums were removed and the famous Corporation Street scheme completed in 1882, a scheme which Joseph Chamberlain declared, "will make Birmingham the richest borough in the country sixty or seventy years hence". The administrative buildings of the city lay until the nineteenth century around St. Martin's church, but here there was no room for modern buildings of a dignified size and design; in 1834 the Assembly Hall, used as a Town Hall, was built at the western end of the eighteenth-century town, and the building nearby of the Council House, Art Gallery, University and the new civic centre has fixed this area as the new administrative quarter.

Birmingham has evolved not from a medieval city but from a little industrialised village; little of the town was laid out on a generous scheme or with any plan at all, much of it was built up very rapidly with streets now considered too narrow, fronted by houses too close together with very few public open spaces. Suburbs were added to suburbs until buildings sprawled over a wide area making approach difficult through traffic-congested streets. It is this evolution which presents modern town planning authorities with great problems; but it also presents them with great opportunities, they are but little hampered by the need to preserve streets and buildings of historic and architectural beauty.

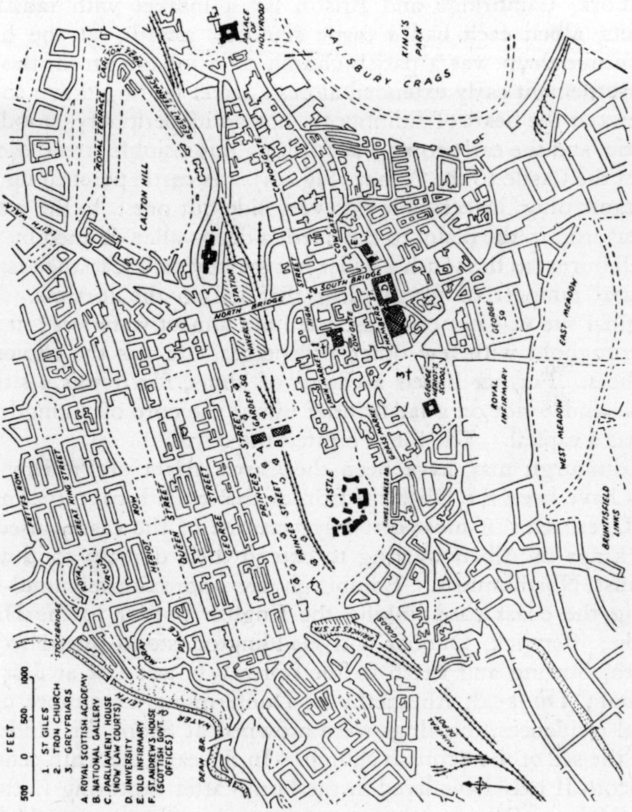

Fig. 33. Town Plan of Edinburgh.

Edinburgh (Fig. 33) the last example, bears clearly the mark of a capital city. Like the other cities discussed it has a long and complex history. It resembles York and Cambridge rather than Bristol and Birmingham both in size and in the way in which its early form still dominates, in spite of recent growth, the character of the city to-day. But the essential plan of the early settlement is different from any considered. At the heart of York, Cambridge and Bristol lay a market with radiating streets, albeit each had a castle standing guard; at the heart of Birmingham was a parish church on a village green, though the settlement early extended along a street leading down to the river; at the heart of Edinburgh, not on its periphery, stood the Castle, and the early town was for long but a single street leading from the Castle to the Abbey (Fig. 34). A castle perched upon a volcanic plug, inaccessible on every side but one; to the west a sheer drop to the plain, to the north a deep valley drained by two small burns, to the south a wider but marshy valley, to the east the natural rampart of glacial drift that led to the fortress. The original site was not only restricted but almost waterless; it was a stronghold with few other advantages. It was the choice of soldiers. Further it was a choice of kings, for it was a site to be defended not on a far frontier but at the core of a kingdom; it was a capital. The kings created the city.

Edinburgh may have been the Castra Alata of Ptolemy; it may have been the Obsessio Etin of the seventh-century Annals of Tigernach; it most certainly was one of the castles used by the kings of Alban warring to extend their domain southward against Northumbria. Edinburgh lay approximately mid-way along the coast road linking the English border to the Highlands; Berwick, Dunbar and Haddington stood guard to the south, Stirling and Perth to the north. It was not at first the capital; Dunkeld, Abernethy or Dunfermline were more usual royal residences, the kings were crowned at Scone, St. Andrews was the see of the Scottish church. In the early eleventh century Malcom II may have lived in the burgh after capturing Lothian, and Malcom III certainly set out from there in 1093 on his last expedition against England. His wife Margaret built the little chapel on the rock which is the oldest surviving building in Edinburgh, and the ferry across the Forth making the easiest link with Dunfermline was already in her day in common use and known as the Passagio Reginae.

It was David I (1124-53) who transformed the fortress. He cleared the castle precincts of all but the soldiery, and established the first settlement outside the castle rock on the natural rampart running east. At the upper end, the townsfolk built their houses, laid out their closes, and held their markets ; at the lower end the Augustinian Canons, who had had a convent within the castle, were granted land at Holyrood. Here they built their abbey, and the lay brethren and workmen engaged strung their houses out along the Canongate. The High Street and the Canongate met at Nether Bow, where from the north came in the road from Leith, from the south the road from the Tweed valley. Around the town a wall of earth and turf was built ; water was caught as it fell from the skies and brought from burns at the castle foot ; beasts were daily driven along the Cowgate to graze upon the Burgh Muir where inhabitants of the rock had rights not only of pasturage but of turf cutting.

In the twelfth century the Border country prospered ; a peaceful invasion took place from the south, Anglo-Norman families were encouraged to settle. Twelve royal burghs and twenty-five monastic houses were established there before the War of Independence of 1296. Charters of Edinburgh of the late twelfth century were addressed to " Francis, Flamingie, Anglis et Scottis " suggesting that the city attracted the foreign element. In the fourteenth century the Border was overrun by English armies and it was two hundred years before order was restored. Edward III's garrison was turned out of Edinburgh in 1341, but Roxburgh was not recaptured by the Scots until 1460 and Berwick after changing hands in 1461 was retaken by the English in 1482 and never regained. The Edinburgh district suffered in these centuries but it gained too. Merchants from Berwick, finding conditions in the Border port too disturbed, came to Leith. The harbour was improved in the fourteenth century and trade increased. Hides and sheepskins, rough cloth, and herrings and salmon were exported in return for timber from the Baltic and wines from Gascony. From Flanders came linen, dried fruits, rice, sugar, ginger, almonds, oil and vinegar reflecting the wide entrepôt trade of Flemish ports. The wider economic links in the neighbourhood profited Edinburgh. In 1482 a royal charter describes Edinburgh as " the most principal borough of our kingdom ", and James IV by the time of his death on Flodden Field (1513) had built his Palace at Holyrood. The defeat at Flodden brought temporary decline ;

but peaceful relations with England were restored, and with peace came renewed prosperity. In 1532 the College of Justices was established and the Supreme Court seated at Edinburgh. Edinburgh was pre-eminent among Scottish burghs: in an assessment of taxation in 1585, the figure for Edinburgh equalled the combined total of the next three, Dundee, Perth and Aberdeen. With the union of the Scottish and English crowns in 1603, Edinburgh no longer had a royal court in permanent residence, but it remained the regular meeting place of the Scottish Parliament and the centre of the law courts. The Reformation strengthened Edinburgh's position; the General Assembly of the Church of Scotland met regularly, and Edinburgh rather than St. Andrews became Scotland's ecclesiastical capital. In 1707 Edinburgh with a population of about 20,000 contributed one-third of the tax levied on all the Scottish boroughs. At that date with the Union of Parliaments Edinburgh ceased to be a legislative capital, but the severance of the administration was never quite complete, and after a period of increasing centralisation in London, the present tendency is towards a greater degree of local independence.

Edinburgh even before the Union was not merely or perhaps primarily an administrative capital, but Union increased, relatively to her other functions, her industrial and commercial importance. The border route instead of being a pathway for reavers became a highway for traders. In the seventeenth century soap, glass and leather industries were added to the earlier woollen and linen manufactures of the city, and in the eighteenth century printing and paper making and coach building became principal industries. Coal was mined and salt evaporated in the vicinity. Union removed restrictions on trade with the colonies and between 1700 and 1800 Leith's shipping increased tenfold. But Edinburgh and South-East Scotland did not in the nineteenth and twentieth centuries remain the centre of gravity of Scotland's economic field. The American trade, especially in tobacco, profited Glasgow more than Leith; and the dislocation of the American trade which followed America's struggle for independence coincided with the great expansion in the iron and steel and textile trades resulting from technical innovations in

Glasgow's immediate hinterland, and thus Glasgow not Edinburgh took the lead in the industrial and commercial life of Scotland. But industries grew in Edinburgh too, she could not have gained and maintained a population of near half a million else. In the twentieth century, paper making, brewing and distilling, printing, flour milling and biscuit making, woollen manufactures, rubber manufactures and structural engineering gave employment to increasing numbers. But industry and commerce never entirely dominated the banking activities, nor the legal, ecclesiastical, administrative and educational interests of this northern capital city.

The structure of Edinburgh can now be examined. The little city of David I with its three distinct quarters, the Castle, the High Street and the Canongate, remains at the core of the modern one. There have of course been great additions to it and alterations within it; its original buildings have disappeared but its design has not been destroyed. In the fourteenth century, after the English garrison was ejected from the Castle, the Constable's Tower on the Castle rock and the Great Stables on its southern flank were built, and the reconstruction of the Norman church of St. Giles begun. To house the workmen employed the first suburb without the walls, Portisburgh, grew up beyond the west gate. In the fifteenth century houses were built along the Cowgate and the Grassmarket, and in the sixteenth parts of the Burgh Muir were built over, and a line of houses appeared along the south side of the Nor loch which had been made by damming sometime in the fourteenth century the two little streams along the valley that is now Princes Street Gardens. The Nor loch was made presumably to strengthen the defences of the Castle. A description written in 1530 by an Edinburgh man resident in Germany gives a picture of Edinburgh at that date:

> "The city itself is not built of brick but of unhewn and square stones and its several houses might be compared with great palaces. In the centre is the Town House (Tolbooth) and the Cathedral church of St. Giles. The palace of the king against the monastery, very large and impressive.

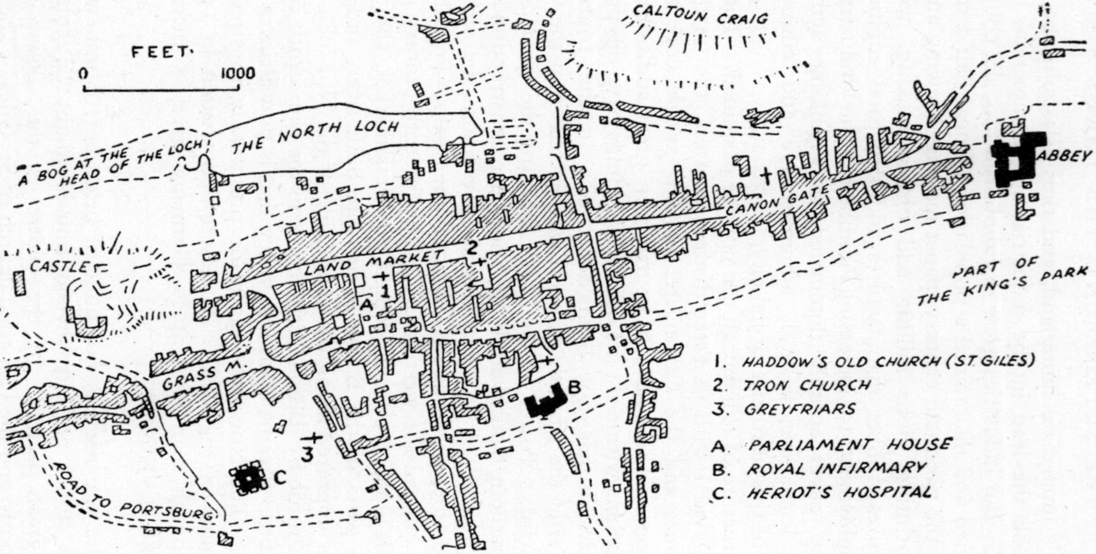

Fig. 34. *A plan of the City and Castle of Edinburgh, by John Rocque.*

On each side of the High Street are famous buildings: from north and south are many little streets all adorned with lofty houses, and in the Cowgate dwell magnates and counsellors of the city and the palaces of the chief men of the kingdom—a street where nothing is homely or rustic but everything magnificent."

Rocque's plan of the middle of the eighteenth century shows little extension in area (Fig. 34). Houses in and around the High Street were built yet more lofty, some had five and even six storeys. Some of these were built round wide open squares, Myles Square opposite Tron Church and the Lawnmarket (originally the Landmarket) contrast with the narrow wynds of the earlier period. To the south of the Cowgate other fine squares were laid out in the eighteenth century. Brown, Argyll and Adams squares were all fashionable quarters until swept away in the nineteenth century, when Chambers Street was built in 1870 parallel to the High Street and the Cowgate to ease movement east to west through the town. Only George Square, south-west of them again, still stands as witness here to the great traditions of eighteenth-century building. But the site was restricted. As early as 1681 plans were first considered for "an Extension of the Royalty or Royal Burgh to open lands beyond the Nor loch". It was a century before the schemes were carried out and the "new town" built to the great design. Between 1770 and 1800 the new quarter was laid out in wide streets and spacious squares, a quarter which remains today in strong contrast both to the complex of courts and wynds between Castle and Palace and to the sprawling suburbs of modern times. Connection had to be established between the New Town and the old; a tenement in the High Street opposite Tron Church was removed and the Nor Loch valley was bridged and the Nor Loch drained. The New Town proved to be not merely an addition; its amenities proved a great attraction which brought about a revolution in the structure of the city. The wealthy citizens moved from their old "lands" on the Royal Mile to houses in New Town squares. Bridge Street was extended south over the Cowgate to link the New Town with the earlier streets and squares laid out to the south, which after the building of the Royal Infirmary in 1738 became the centre of the city's University life. In 1755 the population of Edinburgh was about 31,000; in 1800 it had increased to 162,000 and the town had almost doubled in area. And it was still to grow, both in planned and unplanned fashion. James Craig's plan for the "New Town" had been complete in itself and did not envisage further growth. Its very completeness made for difficulties. The land to the north and east was largely owned by the town and Heriot's Hospital and the next official extension came here. In 1804-6 Robert Reid planned and laid out the region between Heriot's Row and Fettes Row, and alterations to the Royal Circus later gave links to Stockbridge, originally an independent village attracting visitors to its mineral springs, but now absorbed in the growing city. In 1814 a plan of William Playfair's which covered the whole area between the Calton Hill and Leith was adopted. The Royal, Regent and Carlton Terraces were laid out to bear witness still to the dignity of the conception. But first hard times halted progress; then the urgency to build quickly and cheaply on any available land imposed by a rapidly growing population, and the coming of the railways, overwhelmed the stately progress of carefully planned and costly schemes. The town was ringed with small properties and on these owners built as and how and when they pleased. All did not build badly; in the prosperous days of the nineteenth century many built villas in wide streets extending the attractive areas of the town. Residential quarters were built too on the town commons; on Easter and Wester Burgh Muir grew Newington and Morningside, and on Crown land still nearer the Castle grew Whitehouse, Bruntisfield and Merchiston. To these new residential areas people moved once again outwards; Princes Street and George Street and other New Town streets and squares, losing their residents, became the commercial and business centre of the town, taking over the functions of the High Street. Properties in the High Street and Canongate became divided into smaller and smaller tenements to house the Edinburgh poor ever thicker on the ground. The coming of the railways brought industrial quarters close into the city; to the north of Calton Hill, printing works, breweries and engineering shops with goods yards to serve them crowd in below the Royal Terrace ; to the west of the Castle Rock lie the industrial districts of Dalry and Gorgie with their railway yards. The railway penetrated through the very heart of the city : the valley of the Nor Lock offered a route and ground still open because marshy between the "Old Town" and the "New Town". At the western end is Princes Street Station, at the eastern end Waverley Station. Modern landscape gardening has however done much to camouflage the stations and the railway ; and Princes Street Gardens, not the railway, fortunately dominate the eye, and set off the Castle Rock which looks across them to the gracious squares of the New Town.

Twentieth-century Edinburgh has added suburb after suburb until houses now reach the flank of the Pentland Hills, hills until lately only within the orbit of Edinburgh's interest because their springs provided its water. But in the heart of the town history and geography have preserved open spaces—the volcanic crags of Calton Hill and the King's Park are still open ground, the Meadows and the Bruntisfield and Duddington Golf courses are remnants of the medieval town meadow and moors, and public parks and the famous Arboretum of the Edinburgh Botanic Gardens lie along the Water of Leith. Edinburgh, despite some unsightly slums and factory areas near its centre that present grave problems to present town planning, has in its historic buildings, its plan, and the busy life of its streets the air and the structure of a capital city long important and still flourishing.

The towns of Britain preserve perhaps even better than villages the oldest elements in their structure, for some of their buildings are more imposing and thus more enduring. But the towns also change more rapidly than the villages, though the greatest changes in their geography are often of recent date. Their geography is then an attractive theme to the historical geographer : first, the material, archaeological, architectural and historical, from which to construct the form and function of the town at significant dates is often available; secondly, the past form and function of the town is so clearly linked with the present that the former must most obviously be studied in any attempt to understand the latter; and thirdly, the rapid changes in the landscape which man can produce where he concentrates his effort are most effectively illustrated.

An appraising geographical eye is more frequently used in walks abroad than at home; and this is particularly true in towns and cities. The site and situation in relation to structure and function catches the attention in a foreign town; perhaps because of its strangeness, perhaps because of the leisure in which it is viewed. Familiar countrysides are recognised to possess a geography; familiar towns but rarely. Awareness of familiar towns is enhanced by comparison and constrast with those less well known.

Professor Gordon Childe uses the term "Urban Revolution", which suggests at first thought the rise to dominance of towns in the nineteenth and twentieth centuries, to describe the complex changes which in sum first freed, about 3000 B.C., certain groups from the direct production of raw materials for food and clothing, and allowed some specialisation of labour. This revolution took place in the Tigris and Euphrates valleys, in Egypt, and along the shores of the Arabian desert and the eastern Mediterranean perhaps before the earliest farming groups arrived in Britain. It is thus not surprising that the south-eastern Mediterranean impresses northern travellers by the age of its "civilisation." Ur and Babylon, Memphis and Alexandria, some of the caravan cities of the desert fringes were wealthy towns when even Greece, Italy and Spain were lands of small villages. In the Middle East and the Mediterranean can be found many examples of both the stability and instability of town life. Some of the earliest cities are now no more than material for the archaeologist's spade and the historian's pen: Ur, Palmyra, Memphis, Cnossus, Mycenae, Troy and Carthage are but splendid ruins, or present settlements no more than villages. But Cairo, Athens, Rome, Marseilles and Malaga survive as great cities with a continuity of site, if not of status. Athens, with the Acropolis, the temples on its summit, the theatre on its flank, the Agora at its foot; the narrow medieval streets with their Byzantine churches around; the spacious modern squares; the busy road leading to its port of Pireaus epitomises the historical geography of the land in which every European geographer must have an abiding interest.

In north and western Europe, towns and cities are less old and the continuity of town life there from Roman times, even in the oldest of them, is debated. Whatever the period of origin, however, to that period the geographer must look back to comprehend the town. Some seem to have survived in medieval form; Aigues Mortes, once a busy port on a distributary of the Rhône, has changed but little since with improvement in land communications its trade was captured by Marseilles; the Cité of Carcassonne preserves still the form of a medieval frontier fortress; Santiago de Compostela still gives the impression of a medieval town with a shrine as the core of its being. Other towns originally built on natural routes at points where men were wont to congregate—the towns of the Italian piedmont, of the Swiss plateau, of the Danube and the Rhine, of Flanders and the Baltic coast—have a situation and a site with a function changing little with passing centuries. They have but grown in size, and in many, the early castle, the medieval bridge, the guildhalls and merchant houses of earlier days, remain among the shops and stores and factories and residential suburbs of modern times. Capital cities like Vienna and Paris and Rome reveal clearly their long continued metropolitan status.

Planned cities are to be found in Europe too, cities laid out to a design essentially of one period. The "bastides" of southern France, towns built, as a rule rectangular in design, to recolonise and hold the frontier lands devastated by French and English armies. Sauveterre-d'Aveyron and Labastide Cezeracq show their origin in name; and many more, Saint-Macaire, Cadillac and Lavardac, for example, show it clearly in their fourteenth-century buildings. The capital cities of Madrid and Berlin differ markedly in situation, site and function, as well as in plan, from Vienna and Paris.

It is particularly in the towns of colonial areas that the "newness" of the civilisation is felt. Even in Boston and

Philadelphia, for long the first cities of the land, life seems young. In Cape Town, in Washington, and still more in Canberra, the future not the past seems to dominate the scene. It is particularly in the towns too that the rapidity of the growth of the society is most strongly sensed. From frontier trading posts to great ports, Cincinnatti, St. Louis, Kansas and Chicago have been transmuted as it were over night; and still greater has been the change from mining camps to wealthy cities of places like San Francisco and Johannesburg. Industrial towns everywhere essentially belong to the late nineteenth or the twentieth century; but in the "new" countries they are more often all of a period, much more rarely obviously grafted on to an earlier town or village than in Europe and Asia. But again it is the time as well as the place of their development that creates their character.

Map Exercises

Suggestions of maps the reader may like to make of his own town.

1. Relief map of built-up area of town.
2. Sub-soil map of built-up area of town.
3. Distribution map of pre-Roman and Roman finds in town area
4. Plan of medieval streets and buildings.
5. Sketch map from earliest extant town plan, e.g. Speed, County town plan
6. [1]Town plan from first edition O.S.
7. Town plan *c.* 1914.
8. Town plan *c.* 1930.
9. Present Town Plan.
10. The communications linking town to surrounding area.
    (a) before the railway age.
    (b) at present.

[1] Maps 6-9 may be effectively combined to show on one map by shading or colours the additions of successive periods.

CHAPTER VI

THE CHANGING GEOGRAPHY OF THE COUNTRYSIDE

I

THE stable elements of the geographical pattern of the countryside, many of which owe their character to conditions prevailing during the early phases of the colonisation and settlement of the land which are discussed in Chapter IV, are but a part of an ever-changing design. Man more or less rapidly transforms the countryside; forest and marsh, heath and moor disappear before the plough, and with time the cultivated land too changes its look, for man does not always work and use it in the same way.

This chapter then makes some attempt to show how to trace the stages in the disappearance of the wild and the development of the sown in Britain, and to illustrate the geographical implications of these changes. It is then suggested that the reader look abroad to Europe, to the New World and to tropical lands where similar changes have been and are being wrought with major consequences for world geography.

II

A. G. Tansley in *Britain's Green Mantle* estimates that before the war (1939-1945) about half of the whole surface of England was covered by grassland of various kinds, rather less than one quarter was under arable crops, about one-eighth was "built up", about one-twentieth was woodland and a little more than one-twentieth was moorland, heath and bog. This distribution of the different uses of land had a well-marked regional pattern, reflecting in general and in detail primarily variations in relief, climate and soil, and in the density of population. Moorland was extensive in the Lake District, the northern Pennines, on Exmoor and Dartmoor; arable land is found dominant in the drier east from Yorkshire through Lincolnshire, the drained Fenland and East Anglia; permanent pasture covered much of the Midlands. In Wales the percentage of moorland and rough pasture was much higher and permanent grassland covered most of the remainder of the surface. In Scotland most of the Highlands and Southern Uplands was moorland, arable land was dominant along the east coast, the coast of the Solway and in the Central Valley. Since rough pasture is here so abundant, cultivated pasture, permanent grassland, was nowhere a common land use. The "plough up" campaign of the war years altered this pattern; much permanent grassland in the pastoral west became arable. In England and Wales the percentage of grassland decreased from 64% in 1939 to 40% in 1944; Scotland could and did change much less.

In contrast to the present scene it is now considered probable that when the Neolithic farmers arrived in Britain the whole island, apart from small areas of waterlogged soil and perhaps a few high and rocky places, was covered by trees. Recent work on the analysis of tree pollen preserved in peat suggests this. The pollen can be identified and thus the make-up of the woods that surrounded the bog or fen at the time any particular layer of peat was accumulating can be determined. The successive peat layers can be approximately dated by their superimposition and their included artifacts, and it can be shown that the climate that followed the retreat of the last ice sheet allowed several successive forest communities to become established before the work of man did much to interfere. The destruction of the last primitive community, a mixed forest, by the axes, ploughs and domesticated animals of man is the greatest transformation of the environment of Britain that has taken place within the historic period.

The Extent of the Woodland: Sources

The extent of the woodland at specific periods in the historic past cannot be precisely determined before the publication of Ordnance Survey sheets. The Ordnance Survey maps give the position of existing woods at the time of their survey, and within the limitations of scale accurately fix the boundaries. For earlier periods use must be made of many and miscellaneous sources to gain whatever information there is.

The reader will learn much from the making of local maps. To consider existing woodlands is an obvious first step. These woods are not of course remnants of virgin forest and are but rarely natural woodlands. It is useful and profitable to try to date the planting of those shown in the latest local sheet of the topographical map of the Ordnance Survey or of the Land Utilisation Survey. Woods planted by the Forestry Commission can usually be easily dated; the planting is within the period of recent local memory. Other woods can be dated by reference to earlier editions of the Ordnance Survey, and to local records, perhaps going back to enclosure award or tithe maps. In some areas where recent changes have been great, either in afforestation or deforestation, it may be profitable to map separately the woods as shown on the earliest edition of the Ordnance Survey, and to compare this nineteenth-century pattern with the present one. An effort should be made to date the planting and cutting down of woods which appear on this map and not on the modern one.

There is no cartographical or statistical evidence that will yield complete geographical detail of the woodland of the seventeenth and eighteenth centuries. This is the more to be regretted since it would seem from the literature that rapid clearance was going on in these centuries. There are, however, two surveys which give an approximate count of trees on land still reckoned as Royal Forests, Parks and Chases: one carried out about 1783, the other in 1608. Both are printed in *House of Commons Journals*, XLVII, 1792. From these sources any forest in the reader's local area may be roughly marked. Exactly where to place the boundaries of the wood and how to distribute the trees within may present difficulties. The precise boundaries of many Royal Forests are not known, and they undoubtedly changed from time to time, and though cultivation and grazing were restricted within the forests, they were not absent; the forests were not stretches of pure woodland, they were remnants of the Kings' hunting grounds "without the law".

Many of the county maps of Saxton, Speed, Morden and others indicate woods and forests by tree symbols or by name but with very varying accuracy and detail: the information on these maps, however, should not be overlooked, and in some areas can be usefully plotted in order to compare the distribution patterns so established with those suggested by other sources.

Though the Domesday record of woodland offers a great many difficulties of interpretation, the mention of wood in the Domesday Book gives a better clue to the distribution of woodland at the end of the eleventh century for the then more fully settled parts of England than can be obtained for any intervening period between then and the present time. Wood and woodland is frequently mentioned in the Survey but the reference to it takes many forms. Sometimes it is recorded just that "there is sufficient wood for the needs of the vill", sometimes that there is "wood for fencing" or for some other particular purpose; in most of the west and southern counties the length and breadth of the wood in leagues is given; and in Lincolnshire the acreage of the wood is stated. In East Anglia the wood is generally recorded in terms of pigs; "there is wood for x pigs" is the commonest entry; but in Sussex, Surrey and Kent the entry gives the number of pigs paid as rent for feeding swine. These entries are difficult to map satisfactorily. It is not known, for example, how much wood is "sufficient". It is not known what is the area of a wood of stated length and breadth; even if the value of "a league" was known or known to be constant from place to place, which it is not, the shape of the wood can only be guessed. It is not known if the phrase "wood for 200 pigs" means that there were 200 pigs in the wood or that 200 pigs could be fed in the wood. Conventional symbols must be adopted and the distribution pattern made must be regarded only as qualitatively, not quantitatively, significant. Even within one county more than one formula may be used to indicate the amount of wood. It is, however, impossible to convert the terms: leagues cannot be changed into acres; acres cannot be changed into pigs, for it is not known how many acres of wood would produce sufficient mast to feed a given number of pigs. This must have varied from place to place with the kind of wood and the condition of growth. Examples of maps of woodland recorded in the Domesday Survey and a discussion of their sources and their interpretation will be found in F. W. Morgan, "Domesday Woodland in South Western England", *Antiquity*, Vol. XVI, 1936, and in H. C. Darby, "Domesday Woodland", *Economic History Review*, 2nd Series, Vol. III, 1950, to mention two articles among many that the reader might find it helpful to consult.

Then much can be learnt of the former extent of local woodland by mapping the place names containing a tree name or an element meaning "wood" or "clearing" (Figs. 15, 16d, 17). The word *wood* itself is common; if the letter "W" is turned up in any gazetteer many examples leap to the eye—Woodbarrow (Som), Woodbridge (Suffolk), Woodhall (Lincs), Woodstock (Oxon). Wood is equally common as a second element: Brentwood (Essex), Goodwood (Sussex), Ringwood (Hants), come instantly to mind. *Wald*, *weald*, originally meant wood, as in Waldingfield (Suffolk), and Waltham (Essex); later it came to mean high forest land, then "open" upland, cleared woodland, as in Cotswolds (Glos), Methwold (Norfolk), Cuxwold (Lincs). *Graf*, a grove, as in Gravesend (Kent) and Bromsgrove (Worcs); *bysc*, a bush, as in Bushey (Herts), *hyrst*, a wooded height, as in Hurstpierpont (Sussex) and Brockenhurst (Hants), suggest perhaps smaller woods. Many elements particularly common in woodland districts refer not to the wood but to the open spaces in the wood; *field* and *ley* in all their various spellings are perhaps the commonest examples. *Feld*, *field* held originally the meaning not of a small area enclosed by hedges or walls but of an open space where all else was closed, i.e. treed. It had something of the same connotation that the Dutch "veldt" still has. In Warwickshire in the eighteenth century the area south of the Avon was still known as the "Feldon" in contra-distinction to the woodland to the north of the river. *Leah*, *leigh*, *ley*, occurs in place names in two senses; sometimes it means wood, as probably where it is joined to a tree name, Ashley (Devon), Oakleigh, (Kent); more often it means an open place in a wood, a glade, as when coupled with animal names, Studley (Oxon), Calverley (Yorks), or with names of crops, Rayleigh (Essex), Flaxley

(Glos), Barley (Herts). A "lea" seems as a rule to have been a smaller open place than a "field". *Thwaite*, which in English dialect often means a forest clearing, may come from the Norwegian "*tveit*", a meadow in a wood. The *-thwaite* names of the north and west like Crosthwaite (Cumb), Bassenthwaite (Cumb), Braithwaite (Cumb), Seathwaite (Lancs), have the same significance as the field and ley names in the south. *Denn*, usually a swine pasture, *-ryding*, cleared or ridded land, *-rod* a clearing, an assart, are other common elements that suggest colonisation of the wood. Many place names contain the names of common trees: Aldershot (Hants), Ashford (Kent), Bexhill (Boxhill) (Sussex), Elmham (Norfolk), Hawthorn (Durham), Linwood (Lime-wood) (Hants), Oakenshaw (Yorks), Thornbrough (Northumb), Willoughby (Lincs), to call to mind a few.

When the local maps have been made, showing the distribution of woodland now, of the woodland of the later half of the nineteenth century, of Royal and other forests known from contemporary descriptions of various dates, of the extent of the Domesday woodland, and of place names containing references to wood these should be examined together to see how the distribution patterns supplement and complement each other. Woods today frequently occur in areas where evidence suggests natural woodland remained late, not often because the natural woodland has survived, nor because physical conditions strongly favoured tree growth there (this would be true everywhere), but because physical conditions did not there strongly attract man to make other use of the land. The maps showing distribution of the woodland should thus be examined alongside the local maps of relief, climate and soil, for they clearly illustrate the principle of the repeated re-emergence of patterns in response to physical factors, provided that technological advance does not change the values. The maps of woodland distribution should also be compared with the maps already made to illustrate the progress of settlement. Hamlet settlement and large parishes have been shown to be characteristic of many areas of relatively late colonisation belonging to the second phase of settlement, i.e. to expansion from sites already established. In conjunction with the evidence of the place names the distribution of the hamlets and the large parishes may help to determine the limits of areas which were wooded until late or lay within the legal "forest".

The Clearance of Woodland

The clearance of the forests and woods of Britain was at first presumably pursued without thought or plan as the immediate needs of time and place suggested; later it may have been given encouragement from authority as it certainly was on the Continent; and later still local or national, imaginary and real shortages of the products of the woods were descried and every effort made to regulate and restrict the cutting down of trees.

The extent of prehistoric clearing is difficult to establish; not even place names descriptive of the land survive in any numbers from this period. But, as we have seen, the distribution of pottery, tools and weapons, of fields and houses that can be dated as pre-Roman suggests that from the light soils of south-eastern England and from the sheltered valleys of highland Britain much of the primitive woodland must have been already cut when Julius Caesar invaded.

Anglo-Saxon place names considered on etymological grounds to be early forms more rarely contain elements meaning wood than those considered to be later forms, therefore place names do not give conclusive evidence of the extent of the wood at the time of the first phase of Anglo-Saxon settlement. Other evidence suggests that the first Anglo-Saxon colonists quite often settled in regions already colonised by their predecessors; it might be taken therefore that the distribution of early Anglo-Saxon names with no element referring to woods in them could be used as a measure of the extent of the clearing at the time they were established. But here as always negative evidence is unreliable if not useless, for it cannot be argued that the place names would necessarily refer to woods even if woodland was there in quantity. Early names in fact seem much more often to be tribal or personal names than geographically descriptive names.

There is on the other hand much evidence to suggest that the expansion of the Anglo-Saxon settlement took place at the expense of wooded land. Place names containing elements for wood are often numerous in areas known from documents to have been forested. Ashford, Sevenoaks, Uckfield, Cuckfield, Midhurst, Petersfield, Haslemere, Chiddingfold, Cranleigh, Crawley, Horleigh, commemorate the colonisation of the great forest of the Weald. Enfield, Hatfield, Braintree, Bocking, Brentwood, Woodford ring Epping Forest. Lichfield, Rugeley, Ashby de la Zouche, Oakham point to settlement in Midland forests. Macclesfield, Ashton, Burnley, Hellifield on the west, and Chesterfield, Sheffield, Barnsley, Huddersfield, Wakefield, Shipley, Bingley, Keighley on the east, bear witness to the creeping of the farmers up the wooded valleys of the Pennines. The place names with elements for wood are frequently though not exclusively associated with place names which suggest secondary settlement—the *-steads*, *-stows*, *-stokes*, *-cotes* and *-wics*, with hamlet settlement, and with large or divided parishes. Many of the villages of the Weald of Kent and Surrey and Sussex, and of the Hampshire Basin, for example, have these characteristics of name, form and area and their distribution may help to define the old areas of Andreadsweald and the New Forest. In the region known still as the Chiltern Forest, around Cannock Chase and in the area of the old Macclesfield Forest hamlets are scattered thickly. In Northamptonshire, for example, large parishes with appropriate names—e.g. Oakley, Brigstock, Benefield, Southwick, Blatherswyke, Bulwich, Apethorpe—help to mark Rockingham Forest. Divided parishes, indicative perhaps of the rapid thickening of colonisation, once begun, of accessible areas along the frontier where settlement had been temporarily held back by natural or man-made obstacles, are characteristic of many old forests. Within a small district of Epping Forest there are Great and Little Dunmow, High Easter and Good Easter, High Laver, Little Laver and Magdalen Laver, and the group of Rodings. Two Rodings are mentioned in a document of 1050, all six in documents of the thirteenth century.

To give an absolute date to the phase of expansion and thus to the period of much woodland clearance is more difficult. The phase was not contemporary in all districts, and was a longer one in some areas than in others. The record of many of the woodland and clearing names, and of other names associated with them in the same area of distribution, in pre-Conquest documents and in the Domesday Survey suggest that colonisation in woodland areas had progressed a long way by the time of the Norman Conquest.

The Domesday Survey for all its inadequacies does enable some broad conclusions to be reached about the extent of woodland at the end of the eleventh century. Most entries give the number of ploughteams kept on the desmesne and by the peasants. Even allowing that one ploughteam might plough more in one place than another because of the character of the relief or the condition of the soil or because one group worked its animals harder than another, in general it may be taken that where there are most ploughs recorded per unit area, there was most cultivated land. On the other hand, where most wood is mentioned, there land is at least not so intensively cultivated. All the facts relating to ploughteams and woodlands given in the Domesday Survey have not yet been plotted, or at least the maps have not yet been published, but enough work has been done to make it certain that H. C. Darby is right when he states, "every page of the Domesday shows that there was much more wood then than now". Four qualifications must be borne in mind. First, as has been already stated, owing to the variety of ways in which wood is recorded and our ignorance of the exact meaning of some of the statements and our inability to make even the roughest guess of their equivalent values, little attempt can be made to compare the variation in the density of trees from place to place. Secondly, it is necessary to remember that only wood that was used and the products of which were liable to tax was recorded. Many parts of England which in all probability were heavily wooded in 1086, much of the Weald and large tracts of the Midlands, for example, would have no wood mentioned in the Domesday Survey, not because there was no wood to make fences or to feed swine there, but because wood was not used there for these purposes. Some of the areas showing little wood on a map of Domesday Woodland may be indeed areas largely cleared by that date, but others may be the areas still most heavily forested of all, areas not yet used or settled. The distribution of the plough-teams should help to distinguish the one from the other. Thirdly wood recorded appears in the place where the user lives, not necessarily where the trees grow. It may well be that the apparent scarcity of wood in the Weald in maps of Domesday records reflects the fact that the woods of the Weald were used by downland manors and so entered to them. Where this occurred the downland will appear on the maps to be more wooded than in fact it was. Fourthly, the " legal " forest, land outside the law, outside the " foris ", preserved as hunting ground for the King, some of it real woodland, some of it only heath and moor, was recorded only incidentally in Domesday Book. There is much detail given of the New Forest and some others are mentioned by name, but the position and extent of the Royal Forests cannot be determined from the Domesday Book, nor exactly from any later records.

The rate at which woods were cleared is very difficult to assess. In the Little Domesday Book containing the part of the Survey that refers to Norfolk, Suffolk and Essex, the number of pigs for which there is wood is given for 1086 and 1066. In many manors the number of pigs given for 1086 is smaller than for 1066. It has been argued that these figures give a measure of rate of the disappearance of the wood. But a decrease in the amount of woodland used to feed pigs may merely mean that here wood was taken for other purposes, enclosed for deer parks or to pasture horses and cattle. The pasturing of cattle in woodland would, however, ultimately cause destruction. As A. G. Tansley points out pigs, though eating acorns and beech nuts, do not eat them all; they tramp many into the ground, thus increasing the chance of germination. Pigs rootling around keep the top soil loose and so young shoots can get through. Grazing cattle, on the other hand, tramp the ground hard, and, what is worse, eat the young shoots that do struggle through and thus gradually decrease the woodland they graze by preventing natural regeneration. Furthermore, it is possible that warfare destroyed livestock rather than woodland; often in the very manors where decrease in the number of pigs might be taken to suggest decrease in the amount of woodland there is also a decrease in the number of ploughteams recorded in 1086 over those of 1066. Warfare in itself may, however, hasten the destruction of woods; armies may burn down trees just to "lay waste" the land or to hunt out enemies, or they may cut them down for timber to build camps, roads and bridges. But it is difficult to believe that change in land use, wanton destruction, or the need for timber and fuel destroyed the woods as rapidly in the eleventh century as the decline in the number of pigs in these twenty years would suggest if the rate of destruction in these few manors were typical of England or even of East Anglia as a whole. England would indeed soon have become a bare if not a barren land.

By the thirteenth century, however, it was evidently realised that woodland was not inexhaustible and that for prosperity there must be a nice balance between the wild and the sown. The Statute of Merton in 1235 allowed lords to enclose—i.e. take in for their own use—woodland and waste provided that "it is acknowledged before the justices that they have as much pasture as suffices for their tenements". At this date it would seem that sufficient pasture was the crux, timber itself was probably still abundant and even in the fifteenth century a would-be poet could sing

> "No lack of timber then was felt or fear'd
> In Albion's happy isle."

By Elizabeth's reign a shortage of timber was certainly feared if not felt, and politicians could make capital of the fear, particularly of the fear of a shortage of good English oak to build good English ships. Acts were passed restricting the cutting of trees: twelve young trees must be left on every acre where woods were cut down (35 H.VIII C.17); oak, beech and ash of more than one foot square at the stub should not be cut for iron making (1 Eliz. C.15); and ships' timber was to be preserved within fourteen miles of navigable water (1 Eliz. C.15). The variety of the prohibitions and the number of Acts concerning them in the following reigns may perhaps be accepted as testimony to the reality of the retreat of woodland, at any rate of some kinds in some districts.

The diarist John Evelyn blamed

> " the disproportionate spreading of village caused through that prodigious havoc made by such as lately professing themselves against root and branch . . . were tempted not only to fell and cut down but utterly to extirpate, demolish and raze as it were all those goodly woods and forests which our ancestors left standing for the ornament and service of their country ".

The iron workers too were blamed and Evelyn wished

> " that the exorbitance and increase of devouring iron mills were looked into, . . . and what if some of them were even removed into another world? The Holy Land of New England (there to build ships, erect saw mills near their noble rivers) for they will else ruin Old England. Twere better to purchase all our iron out of America, than thus to exhaust our woods at home, although (I doubt not) they might be so ordered as to be rather a means of conserving them ".

It was of course not only the increase of arable but also the increase of flocks and herds; not only the demand for fuel for iron works but also for tanning, dyeing, baking, brewing and, perhaps the biggest demand, for household fires; not only the cutting of big timber to build ships but also to build or at any rate to roof castles, churches, houses, barns, cattle sheds, sheep folds and pig sties, and to construct ploughs and harrows, vats and bins, carts and waggons: it was all these things that destroyed the woods. In south-east England where men were settled most thickly, near navigable water where timber once cut could be transported most easily, and in south Scotland where though men were fewer trees were fewer too, the virgin forest disappeared earliest. In the Midland Plains the natural woodlands of the deep clay soils were thick and extensive at a much later date. The woods lay on or near the divides between the big rivers far from navigable water and the density of population was not great; there was still plenty of timber here in the seventeenth century. Brick replaces timber in the south where building stone is scarce much earlier than in the Midlands. Coal replaces charcoal for fuel away from the coalfields but near navigable water earlier in the south-east than in the north and west. In the Highlands of Scotland virgin forest still clothes much of the ground even in the eighteenth century; oak woods in the lower valleys, birch and pine on the higher slopes. Imported iron ore was smelted by imported labour on the shores of the sea lochs of Western Scotland in the late eighteenth century because there was abundant wood for the taking to make charcoal. The extension of sheep farming, the preservation of grouse moors, and deer "forests" in the nineteenth century, and the work of the Forestry Commissioners in the twentieth century have slowly but relentlessly now destroyed the natural woodland here too.

The disappearance of the woodland was nowhere of course complete, and not in progress steadily everywhere. In some parts of the country the uplands, that were early cleared by cultivation and grazing and were later abandoned as heavier soils claimed attention, were, where soils were suitable, recolonised by trees. These woods of natural regeneration were as a rule of a light kind, of birch, hazel, thorn and in wetter places of hazel and willow, rather than of oak, ash and elm. Then in the eighteenth century "planting" became fashionable and a few of these plantations, usually well-wooded parklands and magnificent avenues, are still conspicuous and most pleasing features of the landscape. Many fell to the axe during the war of 1914-18, and many of those that escaped then were cut down between 1939-45. Planting, however, even in a settled age had but a limited appeal: oaks and other forest trees mature too slowly for most people to see "a return for money". It was not until war drove home the disadvantages of a shortage of home-grown timber even in the modern age of world-wide commerce that afforestation on a national scale was begun. The serried rows of spruce and fir planted by the Forestry Commission are as obviously a part of the " cultivated " landscape as fields of wheat and barley, turnips and sugar beet, or hay meadows and grassland leys.

The Draining of the Marshland

If the clearance of woodland wrought the most widespread transformation in the landscape, the draining of bog and fen brought about locally the most dramatic changes. Bog and fen restricted man's use of the land more fully and much longer than woodland did. On maps showing the distribution of early place names or of Domesday vills or ploughteams, regions soft with surface water usually stand out as conspicuously blank. There was apparently little settlement by the end of the eleventh century in the lower valley of the Trent, the Isle of Axeholme, the East Anglian Fenland, the Norfolk Broadland, the Romney and Pevensey Marshes and the Somerset Levels. Villages were established around the edges, along the silt bands, and here and there where there was dry ground on islands in their midst, but reclamation of marshland had not apparently proceeded as far as the clearing of the woodland. Draining to be successfully carried out needed careful organisation, considerable engineering skill and, to secure both, heavy capital investment, which in early times could not be easily provided. Then, bog and fenland, like uncleared woodland, was far from being useless ; it yielded as has been shown fish and fowl, reeds and rushes, turf and peat. The value of the marshland products, as well as the difficulties of reclamation, delayed extensive drainage schemes.

Readers who live in or near any of the old fens or mosses may wish to make for themselves a study of the evolution of their geography. Investigations can be made on very much the same lines as have been suggested for woodland areas. Here again it may be profitable to map first the modern remnants of the bog and fen as they appear both on the most recent edition, and on the earliest edition, of the Ordnance Survey sheets. Earlier maps may be consulted with profit if the limitations in accuracy of the county maps of the seventeenth, eighteenth

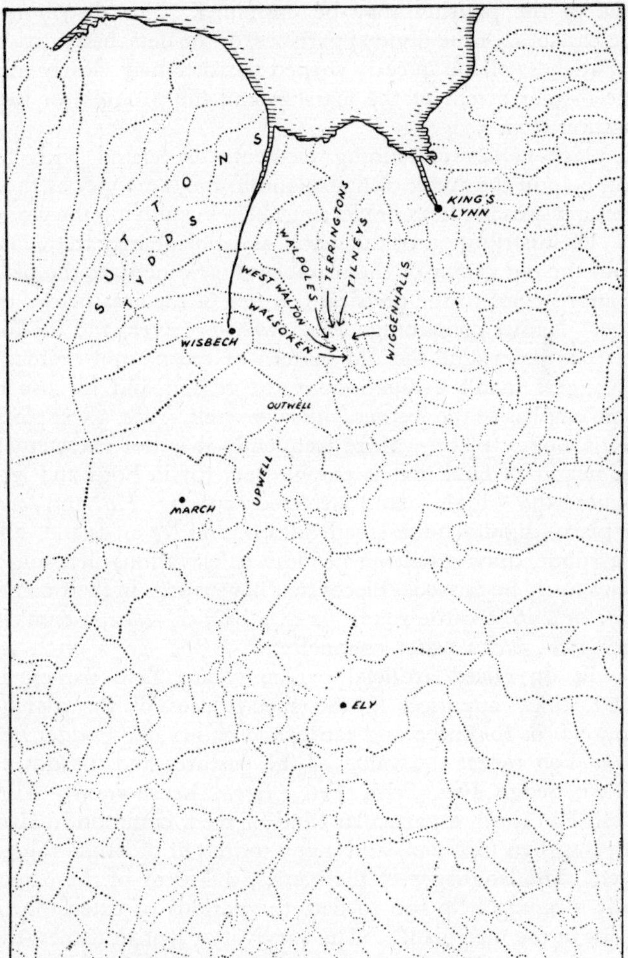

Fig. 35. *The Parishes of the Fenland and the surrounding upland.*

and even nineteenth century are remembered, and the shape and size of the parishes may be examined. The distribution of large parishes, of the divided parishes, of the detached portions of the parishes, of the queerly shaped parishes may clearly indicate the former extent of the marshes and the progress of their colonisation (Fig. 35).

The place names containing elements associated with fen and bog help in the study of marshland in the same way as those containing words for trees or clearings help in studying the woodlands. The distribution of the whole assemblage may frequently mark out the old extent of the marshland. Sometimes the place names afford clues to the physical features of the area before the drainage. Names containing *-lake*, *-mere*, *-ea* (river) and the like may help to determine the old pattern of the natural drainage. Many suggest man's evaluation of the region and his use of it : some emphasise the wetness, like *-fen* itself, *-galle*, a wet place, or *-slade* a boggy place. More frequently it is not the wetness but the height of land that is emphasised, for in bogs and fens if anywhere the " high " land attracted settlers. *Hill* and *-holi*, a high place, dignify quite small bumps, and *-eg* an island, and *-snote*, a snout, draw attention proudly to elevations that might otherwise easily be missed. Places that have words in their names like *draft* or *drift*, a cattle road, *-barre*, a barrier, *-cluse*, a dam, or *-ea*, *-lad*, *-gole- gote*, a water channel and *-delph*, *-dic*, a ditch are often built on raised artificial banks. Many field names, in particular, draw attention to the special uses of the marsh : the many words for sheep and cattle, and names like Fodder Fen and Mow Fen reflect the value of the pastures and meadows ; Reed Fen, Sedge Fen, Frith Fen (*fyrhp*, brushwood), Turf Fen, Lesh Fen (lesh, a particular kind of reed, common in East Anglia), Lug Fen (*lug*, flax) and many more tell of other valued products. The languages of the names, the form of the names and their meaning help too to date the periods of extension of settlement in the marshland. The survival of British names, the scarcity of Anglo-Saxon names, the number of Scandinavian names, names associated with particular individuals like Morton's Leam, Bedford Level, Wellington Farm, all afford dating evidence. Thus the assemblage of the names can be used not only to map out the area of the marshland, but also to trace the progress of its reclamation.

Most fens existed in approximately their natural state until so late a stage of the settlement that there are usually numerous and good written records describing their former extent, their natural products, their drainage and settlement. Local records, maps, rent rolls, deeds, taxation returns, acts, and descriptions should all be searched for information.

A brief outline of the changing geography of the East Anglian Fenland will serve to show the variety of local sources that can be used to establish the facts about each successive scene, and to emphasise the spectacular nature of the transformation that man can bring about locally as his technological knowledge and equipment increases. The change may be so great that completely new estimates are made of geographical values. It will illustrate too the close interplay between the changes wrought by nature and those wrought by man, and serve to bring home the fact that man intending one result often provokes another unforeseen and by no means always welcome.

Air photographs reveal beneath the present pattern of buildings, drains, roads and fields, other constructions and other fields now abandoned. The archaeologists' spades turn up finds of Neolithic and Bronze Age date, to show that the area was by no means an empty unused waste in prehistoric times. Finds of the Roman period suggest that cultivation accompanied by draining and embanking was going on in parts of the area then; work to discover how far Roman reclamation and settlement extended is at present in progress. Perhaps the early efforts of the Romans engendered later failure; the chain of cause and effect—drainage, shrinkage, wastage, silting, drowning—that harassed their successors may also have bothered them. Perhaps Nature played in addition an independent hand; there is evidence in silt-covered Roman sites on and near the coasts both here and in the Low Countries of a relative rise of sea level in late Roman times which would have caused widespread flooding of the low-lying area. Certainly, there is less evidence of settlement in the Fenland in early Anglo-Saxon times than in the Roman and in some prehistoric periods. This may be the

Fig. 36. The Medieval Fenland.

result of the destruction of material remains in sites continuously occupied since the Anglo-Saxon settlement, it may indeed be the confirmation of the oft-repeated tale that Anglo-Saxons feared and shunned the marsh, but it may well be that the physical conditions in the area worsened in the third century and lessened the odds that men would survive to build flourishing societies. In the south the islands were colonised: the foundation of the abbeys at Ely and Ramsey, Whittlesey and Thorney show monks in their traditional role as " seekers after solitude ", as " colonisers of the waste ". In the north, on slightly drier land between the extensive freshwater marshes inland and the salt marshes of the Wash, a line of villages was established behind the great sea bank of Anglo-Saxon or perhaps earlier date that runs from the estuary at Lynn to the estuary at Wisbech and beyond: Terrington, the *-tun* of Tira's people, an early name form, Tilney, Tila's *eg*, recording the island character of the site, and Walpole (the pool by the wall), Walton and Walsoken, referring in name to the all-important wall (Fig. 36).

But the marsh itself was empty; empty but not useless. The run of the parish boundaries of the settlements around the edges and on the islands shows the care taken to include within the lands of each village its share of fen. The complicated pattern of detached portions in the Norfolk marshland (Fig. 35) probably reflects the old rights of intercommoning, and the survival of the pattern is a testimony to the value placed upon these rights. The Domesday record of fisheries and salt pans belonging to manors within easy reach of the Fen, which must in fact have been situated within the Fen, gives direct evidence of the value of two important Fenland products. The township assessment of Norfolk, 1334 (Fig. 37), shows the villages of the Marshland assessed at higher rates than any others, they were big villages, the centre of large parishes, it is true, but the value per acre of the land here was above average and yet much of the area was marsh. Anyone who knows the great churches of the Marshland of Norfolk, Cambridgeshire and Lincoln needs no persuading that wealth was to be gained from the fen long before the extensive drainage schemes of the seventeenth and eighteenth centuries. The magnificent late Norman of

Walsoken, the beautiful Early English of West Walton point to the wealth of the area around Wisbech before the diversion of the drainage, and with it much of the trade, along the Well Creek to King's Lynn. The lovely churches of Wiggenhall St. Mary Magdalen with its glass, Wiggenhall St. Mary the Virgin

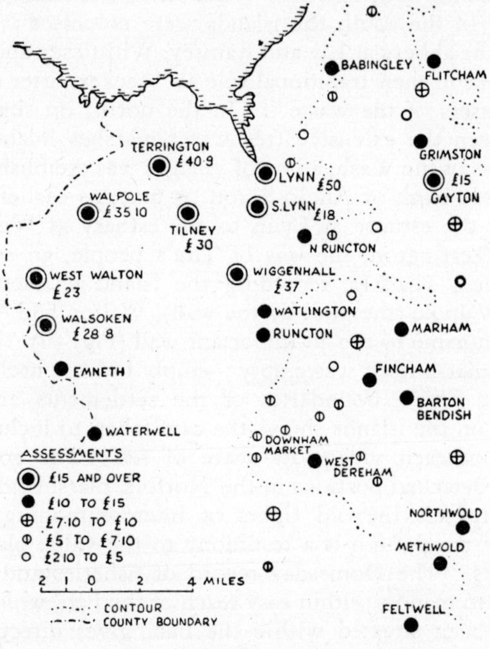

Fig. 37. The Value of the Townships of Western Norfolk in 1334.

and Wiggenhall St. Germaine with their pews, Tilney All Saints with its great late Norman nave and chancel, Terrington St. John with its detached tower, Terrington St. Clement, the largest and grandest, and Walpole St. Peter, perhaps the most lovely of them all, bear witness not only to the piety but also to the prosperity of their builders. It is often said that it was wealth from wool that built the churches; marshland sheep

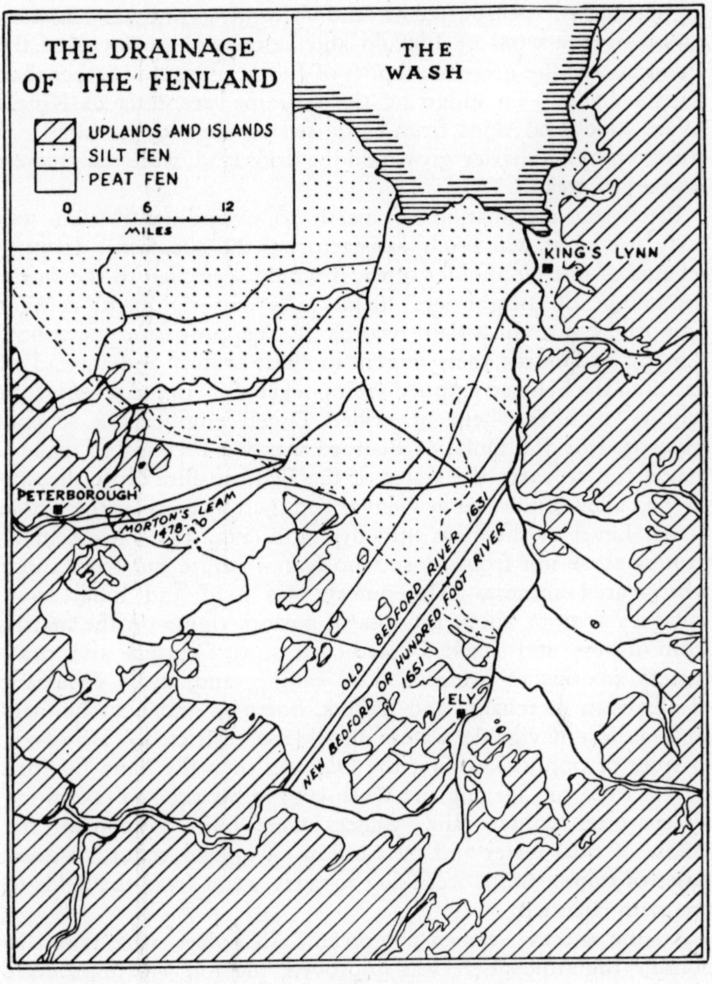

Fig. 38. The Drainage of the Fenland.

were numerous doubtless but Norfolk wool did not command a high price in the fourteenth and fifteenth centuries. Boston, exporting the wool of Lincoln and Leicestershire flocks of the uplands, was the great wool port of Eastern England rather than King's Lynn. To judge by the customs' accounts of King's Lynn, hides and skins from cattle fed on the rich pastures and corn, wheat and barley grown on the drier land, were the greatest sources of wealth.

Piecemeal reclamation of marsh to extend cultivation was early tried but was rarely satisfactory; the successful draining scheme of one landholder often, it would seem, led to the worse drowning of neighbours' fields. As an eighteenth-century Cambridgeshire historian records in inappropriate metaphor, "many had burnt their fingers in these waters and instead of draining the Fen had emptied their pockets". In the seventeenth century with the help of skilled Dutch engineers a scheme sponsored by the Duke of Bedford was put forward to make a major cut, to carry the water of the Bedfordshire Ouse directly to the valley of the Nar at Denver and thereby convert the whole South Level of the Fens into dry land (Fig. 38). A storm of protest arose not from those who feared failure but from those who feared success. The outcry was loud and long, from villagers anxious about the loss of pasture rights on the marsh, from fishers and fowlers, reed-cutters, turf-cutters and peat-cutters anxious to preserve their calling, and, most vehement of all, from merchants and traders, boatmen and boat builders anxious lest navigable waters should be restricted. However, in spite of opposition, first the Old and then the New Bedford Levels were cut, leaving the Wash-land in between to take flood waters after winter rain. Sluices were built to control the inflow of tidal water and the outflow of upland water; a great series of minor drains were dug to carry water from the fields to the main channels.

At first success seemed complete. All voices were raised in praise: the atmosphere was improved, the soil was improved; fine crops were raised of wheat, oats and barley, peas and beans, turnips and onions, coleseed, chicory, mustard and woad. Thomas Fuller in his *Church History* of 1655 writes:

> " the chiefest complaint I hear of is this, that the country hereabout is now subject to a great new drowning, even to a deluge, an inundation of plenty, all commodities being grown too cheap therein ".

But this blissful state of affairs was not to last, already by the end of the seventeenth century it was evident that the waters had not been permanently defeated. Celia Fiennes riding from Bury to Peterborough by Newmarket and Ely in 1698 describes floods around Ely and gives no hint that they are unusual. Defoe, descending what he calls the Hogmagog Hills to Cambridge, writes in 1724:

> " We saw the Fen country on our right almost all covered with water like a sea. The Michaelmas rains having been very great that year sent down vast floods of water from Upland countries; and those Fens being the sink of no less than 12 Counties are often thus overflowed. . . ."

Inherent in success lay failure; the story of the changing geography of the Fenland illustrates excellently the complexity of the inter-relationships between the elements that in balance determine the character of the landscape, and emphasises the necessity for an understanding of this complexity if the far-reaching long-term changes that interference will bring about are to be the changes anticipated and desired.

As the fen was drained the level sank, since both peat and silt had less volume dry than wet. This in itself reduced the natural fall of the land towards the sea, already low enough if rivers and drains were to run freely. Then strong winds are common in the area, there were no trees to shelter the fields, fine top soil blew away and further lowered the land. The better the drainage, the less the outfall, the more difficult continued drainage. But worse: the Fenland lies in an open basin, the inner part filled with peat, the seaward part with silt (Fig. 38). The peat when drained not only shrank but was attacked by soil bacteria which, unable to live when the soil pores were full of water, now flourished in the soil air and literally "ate" the peat away. The silt, largely made up of mineral particles of silica and alumina, did not suffer the same fate, therefore the peat area inland gradually became lower than the silt area towards the sea, and then what free flow of water there was, was in the wrong direction. Furthermore, water in the drains deposited silt along their courses, building up their beds and their banks while the peat beneath this local deposit of silt did not shrink or waste. Thus the drains became higher than the surrounding land, and the big drains deposited more and built up their beds and banks more quickly than the small drains, so that the main drains became higher than the lesser drains and the field drains which had been cut to feed them. In addition the Wash was a shallow and confined gulf with cross winds and wave action constantly shifting sandbanks; the more water that was drained off the land, the more silt was deposited near the outfall since the river current was not strong enough to carry it far out to sea, the bigger and more unstable grew the sandbanks, the more the mouths of the rivers were obstructed, the less became the gradient of the drains.

It was soon obvious that pumping was the only solution. But it is also clear that the more effective the pumping and the more effective the drainage the greater the difficulties to be overcome. Descriptions of the Fenland continue to give period by period in alternating succession pictures of great and mounting prosperity, and then of floods, devastation and poverty. At first little pumps worked by donkeys were tried but the donkeys could not keep pace with the shrinkage and wastage of the peat. Then windmills were built, the whole Fenland was divided up into small areas, each with its own pumping system, a great series of windmills turning "scoop wheels" ladled water from lesser and therefore lower drains to greater and therefore higher drains, and along the main drains from lock to lock and thus to the outfall. By 1748 there were 250 windmills turning in the wind, in the Middle Fen alone, and once more in 1753 a local writer describes the area as "much improved by drainage. . . . The fens after a dry or not overwet winter, produce abundance of oats or coleseed; as also great plenty of very good turf, with fish and fowl to admiration ". All seemed well in the Fenland and the natural products of the marsh, in spite of fears, were not ousted yet, at any rate, by the new crops growing abundantly on the drained land.

However numerous, the windmills were at the mercy of gales, calm and frost, and the more water pumped the more need to pump. The more efficient the windmills the quicker the land level sank, the greater the " lift " needed; the smaller the gradient, the more the outfall silted, the greater the sandbanks grew. Disaster again overtook the Fens. Arthur Young, the first Secretary to the newly founded Board of Agriculture writes in 1805, " many fens all waste and water " where twenty years previously " buildings, farmers and cultivation ". The innovation of steam pumps once more changed the scene: in 1827 people are still " fearful of entering the Fens lest the marsh miasma shortens their lives ", by 1838 Joseph Glyn, one of the pioneers of steam pumping, " had pleasure to see crop plants of wheat and oats again take the place of sedge and bulrush ", and in 1854 Henry Gunning, a Fellow of Christ's College, Cambridge, rejoiced that as " sportsmen from the University ", no longer indulge a passion for shooting in the fens, he is " happy to say that this incentive to idleness no longer exists ". Steam pumping has indeed changed conditions of life in the Fens, but even today man has not achieved a complete control of Nature as those who read about, saw or still more suffered in the floods of 1947 will fully realise.

Reclamation of Moor, Heath and Downland

Reclamation of moor, heath and downland is less spectacular because less complete. For the most part moorland is not strictly speaking " reclaimed " at all; it is used more or less in its natural state for rough grazing. Animals, mostly sheep, grazing there, though they have modified the natural vegetation, have not transformed it. To some extent this is also true of the downland and the heaths that, on the poor limey and sandy soils respectively, are the nearest remnant of the wild in the South and East. These heaths and downlands may well have been transformed, even created (Chap. IV), by early man, but if his earlier self did much to make these landscapes his later self has recently much modified them. Seeded pastures and even arable fields appear now on parts of the downland that have not known the plough since the days of the Anglo-Saxon invasions. The market gardeners and speculative builders with their preference for light sandy soil have in the last hundred years made great inroads on ancient heaths, and the introduction of root crops which were then fed to folded sheep made the cultivation of much hitherto heathy land profitable in the eighteenth century. Gregory King estimates in 1695 that 25% of land in England was uncultivated Mountain, Moor and Heath, in 1939 it was estimated that about 5% of land in England and Wales was heath and moor. Even allowing for the probable difference in accuracy between the two estimates it is perfectly clear that a great geographical change has taken place between the two periods.

The wild has very nearly disappeared in Britain but just enough is left here and there to suggest the former scene and to call to mind the changes the landscape has undergone. The mountain areas of Scotland, the Lake District, the Pennines, Wales and even Dartmoor have been changed little by the hand of man; there are still considerable stretches of heath, for example, about Heathfield and Hindhead, behind Cromer and Southwold, around Thetford and Sandringham, the latter with something of the character still of a Royal Forest or Chase; there is much downland in the Isle of Purbeck, on Salisbury Plain and, though diminishing, there is still some along the North and South Downs. The Forests are but shades of their former selves; the New Forest has still its feel, Sherwood Forest and Epping Forest have a semblance of tree-covered space but the Forest of Arden and Charnwood Forest have all but gone. Even in the biggest towns a scrap of the wild is sometimes preserved, even the two capital cities have a few remnants: Edinburgh has the King's Park and Calton Hill; in London there is Hampstead Heath though Cricklewood has gone, there is still heath of a sort at Blackheath though no trees now crown Forest Hill, even part of Hackney Marsh and Hackney Downs remain but the words Shepherd's Bush and Wormwood Scrubs now bring first to mind scenes quite unconnected with the vegetation of the land.

The New Farming

Major changes in the geography of Britain's countryside have been brought about since the early days of colonisation not only by an extension of cultivation but also by changes in cultivation.

The cultivated landscape for long changed little; some areas were always progressive, some areas always individual, but it was not until the eighteenth century that there were signs of a general change comparable in scale to that brought about by the laying-out of farms by Anglo-Saxons in the English lowlands. The change was not sudden; after about a millennium following one fashion, it took some two centuries to discard the old ways and take up the new. It was, as in Anglo-Saxon days, a change in techniques which brought about change in geographical values and thus change in the geographical scene. First the introduction of new implements made the light soils much valued for the first time since the introduction of the mould board plough; they were sought again as a first, not a second best, choice of arable farmers. Secondly, stock rearing and stock breeding for its own sake became of relatively greater importance; not perhaps since the days of the Beaker folk had pasture played so large a part in farming economy as it did at the end of the nineteenth century. Thirdly, the unit of farming was reorganised, the little square fields of the Celtic farmers are recalled as eighteenth- and nineteenth-century hedges criss-cross the countryside that had so long lain bare and open in common fields.

The reader should try to trace the changes in an area well known to him and to note the local conditions that affected the time and place of their occurrence. Contemporary literature describing the agricultural landscape is abundant, and, for the latter half of the nineteenth century, there is comprehensive statistical material too, for the Board of Agriculture has published annual returns since 1866. The series of reports drawn up at the time of the foundation of the Board of Agriculture in 1793 give for the first time the means of obtaining a comprehensive and reliable picture of the agricultural geography of England. Everyone is strongly advised to read the *General View of the Agriculture of the County of* ——— that describes his own area, and then to read another volume describing a county that he also knows well but which differs in position, climate, soil and perhaps economy from his own. There are two editions of most of these reports, the first published in 1795-96, the second revised edition published in 1806-10. It is often most instructive to compare the two editions of any one region, and to compare both with the corresponding volume of William Marshall, *General Survey of the Rural Economy of England*, published in 12 volumes between 1787 and 1798. If a few areas are carefully selected, it will be clear that there was considerable variation in farming practice from place to place, a fact not always sufficiently emphasised in the textbooks. Much general and local information for the eighteenth century can be culled from Defoe's *Tour of Great Britain*, and from the letters and journals of Arthur Young. For the nineteenth century none can afford to neglect William Cobbett's *Rural Rides* (1821-1833) and James Caird's *English Agriculture* 1850-51. The geographer touring the country with Defoe, Young, Cobbett, or Caird should take notes *on* a fair-sized map; in this way the important variations from region to region will be emphasised.

New Crops and New Methods

The light soils claimed attention in the eighteenth century not at first for their value but because they offered most opportunity for improvement. Yields may have been falling by now on the heavier loams and clays cropped for centuries with only a triennial fallow to restore fertility, but these lands were apparently still in relatively good heart and practices here were well known and thought well proven. If more was wanted from land, and it most certainly was, then either new land must be brought into cultivation or poorly cultivated land must be better used. But in much of Lowland Britain heath and down on sands and chalk was often all there was left to reclaim from the wild, and the cultivated fields most neglected and unprofitable were on light soils too. New and better ways of farming the light land were needed.

Jethro Tull (1674-1741) published his book *The New Horse-hoeing Husbandry* in 1733. He had a queer idea that plants could only absorb as food small particles so he recommended frequent hoeing during the growth of the crop as well as frequent ploughing before sowing. Though his reasoning was wrong, this part at least of his practice was good and was particularly helpful on thin soils. Frequent cultivation maintained free passage of water in the soil and kept down weeds, and thus increased the amount of food available for the crop. The horse hoe, however, could only be used to advantage if the seed was planted in evenly spaced rows instead of sown broadcast. So with the horse hoe came the seed drill. The seed drill was in itself an advance since it saved seed. It is true that if the seed rows were as far apart as early enthusiasts recommended then land was wasted, but this waste was least on thin soils, for light land would not carry thick crops however sown. If land could be thoroughly cultivated and weeded while a crop was growing, it was obviously uneconomic to leave it fallow merely to clean it. Grain crops could be hoed in the early stages of growth but it would obviously damage these crops to hoe them later on. Root crops, turnips and potatoes, on the other hand, were low growing thoughout, and further, the essential part of the crop was underground safe from horses' hoofs. Potatoes and turnips were not only good cleaning crops but they were shallower rooted than cereals and drew most nourishment from a different layer of the soil and perhaps also preferred a slightly different diet; the soil could then not only be cleaned but, in part, rested by a root crop. If a crop could be substituted in the rotation for the fallow without harm then continuous cropping would to some extent compensate on light land for the relatively low yield from each crop. But these light lands could not stand an alternate rotation of corn and roots. A grass crop of sown ryegrass or clover was introduced; it could be cut probably twice in the summer for hay and then in autumn it was ploughed in to feed the soil; if it was clover, it would also especially replenish the store of nitrogen.

Good though the rotation might be, plant food would soon have been exhausted on land continuously cropped without careful manuring. The introduction of root crops and of sown grass for hay enabled, however, far more beasts to be kept, since hay and turnips provided *winter* food. Turnips were usually fed to sheep on the field, the sheep were systematically folded so as to spread their droppings evenly; the tramp of their small feet may, in addition, have helped fertility by consolidating a little the light land. Cattle were usually yard or stall fed; they were bedded down with straw, and the muck and straw carted to and spread over the fields. Since the great autumn slaughter was now no longer inevitable it became worth while to breed better animals. Experiments in animal breeding, empirical though they were, were successful. Robert Bakewell (1725-95) at Dishley bred his famous Leicester sheep and longhorn cattle. Soon the all-purpose rather runty little animals of medieval England were largely replaced by specialised breeds: of sheep famous for their wool or mutton, of cattle famous as "milkers" or "store beeves", of horses bred for the plough or the carriage.

It was under the New Husbandry, as these practices came to be called, that the Good Sands Region of Norfolk gained and earned its name. Lord Townshend (1674-1738) by the success of his farming at Raynham popularised, as his nickname Turnip Townshend commemorates, the use of the turnip as a field crop, and he established the general lines of progressive farming in Norfolk. Thomas William Coke (1752-1824), 1st Earl of Leicester came to live on his ancestral estates at Holkham in 1776 and he did much to develop further Norfolk agriculture. By encouraging visitors at his house-parties, harvest festivals and sheep shearings, he also did much to spread the knowledge of Norfolk practices throughout the land. Many farmers on Lincoln heaths, Northamptonshire sands and in the drier parts of Eastern England in general, took up the new methods and enjoyed a wholly new prosperity. Everyone was not progressive, old ways of doing things disappeared slowly even in Eastern England, but the New Husbandry by the end of the eighteenth century had wrought great geographical changes. Farming in the new way had proved more profitable than in the old, and the new ways were easiest, were at first only possible, on light well-drained land. The light lands therefore had changed in value, instead of being despised and rejected, they were sought after and prized. On some light lands were now to be found the best farm land in the country.

Though the new farming favoured the naturally well-drained soils, there were not enough of these to satisfy the need for food of the rapidly growing population. So England's progressive farmers looked down the hill and to the north and west. The desire or the need to extend the use of the new methods and new crops once more focused attention on the age-old problem of effective drainage of the clay lands. For fifteen hundred years there had been little change in draining practice; the double furrows between the ridges served as field drains, and only by varying the width and the height of the ridges was their capacity to carry off surface water adjusted to the need. The new machines, the seed drills, the horse hoes, and the reapers when they came (Patrick Bell of Angus invented one in 1828), were difficult to use on high-backed lands. The new crops, turnips and potatoes would not do well on ill-drained soil. Efficient drains must be constructed, but drains that would not hinder the movement of man and beast and machine on the surface; they must be underground. James Smith (1789-1850) of Perthshire in his book, *Remarks on Thorough Drainage and Deep Ploughing*, published in 1835 advocated cutting drains 16 ft. to 21 ft. apart and 2½ ft. deep and filling them in with sticks and stones (he had, he tells, plenty of stones on his surface to get rid of) before replacing the soil on top. Hollow tile drains were next invented and widely used. In the heavy clay lands of South-East England, in Essex in particular, a method of draining that depended upon the cohesive property of the clay sub-soil was introduced. A plug was drawn through the sub-soil, at first by the laborious method of cutting a deep drain, fitting in the plug, packing it over with clay then top soil, and then drawing out the plug. Mole ploughs came into use about the middle of the century, heavy cumbersome things which by various forms of winch and cable tackle drew the plug along the course of the drain without first cutting a ditch.

Effective and controlled drainage allowed in turn the more effective and economical use of manure ; excess water no longer washed away valuable organic and mineral elements as soon as they were added. Sheep could be folded on the land without suffering from foot rot and liver fluke. Cattle grazed upon pastures without poaching them, animal manure thus became more abundant. For better or worse artificial manures came within the farmers' ken in the mid-nineteenth century. In 1830 the first consignment of nitrate of soda was imported from Chile, and in 1843 the manufacture of superphosphates was begun. This is not the place to raise the question of the ultimate effects on land, plants, animals and men of the use of artificial manures. The use of artificial manures undoubtedly, which is the point of interest here, temporarily at least increased yields and cheapened production on many soils.

Mole and tile drainage and abundant cheap manure now made it possible to farm the soils of the rainy west and the clay lands of the east in the new fashion of the light lands. In many parts of the Central and Western Midlands the new rotations were taken up and, where turnips did not do well on the heavier soils, mangolds were used instead. In some parts of the west and north, in Cornwall and Somerset, in Lancashire and Cheshire, in Cumberland and South Scotland, potatoes were often grown on lighter lands instead of turnips. In much of the north and west long leys still succeeded grain crops, but these leys were now often sown with selected grass and clover seed and arable temporarily outworn was no longer allowed just to " tumble down " to pasture.

But times were changing : the stimulus of war-time production passed, the post-war boom was over, industry was attracting labour from the land, and cereals grown cheaply on new land in distant parts of the world entered the home market. Arable farming with wheat production as its main end was becoming a precarious and unattractive investment for capital and labour. More and more of the wetter lands and the heavier lands, in spite of the technological improvements which would have made arable farming on the most progressive lines possible on them, were being laid down to grass because arable farming was too costly to pay there. The second change, the increase of grassland farming, thus prevented the heavier soils from coming fully into their own alongside the lighter ones as arable land farmed in the new way.

Pastures

The emphasis on pasture land was fostered by changing geographical conditions within Britain itself and in the world as a whole.

Within Britain not only were people becoming thicker and thicker upon the ground and thus demanding more and more food, but the relative number of people living by farming and living by manufacturing and trading was changing. In the census of 1911 for the first time those engaged in agriculture do not form the largest single occupational group. This in itself brought about a change in demand and a change in the distribution of that demand. The town dwellers demanded *cheap* food and they did not care how or where it was grown; they also demanded large quantities of food in fresh condition in a few places. Grain would stand a long journey and storage without deterioration. Meat, as living beasts, could walk to market from a considerable distance, but milk, vegetables and fruit must be produced near at hand.

But Britain is not " a world unto itself ". Pioneer farmers, armed now with ploughs with steel coulters and shares able to cut and furrow the rough prairie sod, were fast colonising the virgin soil of America's " New North West ". There, where high yields could be obtained without extensive clearing, draining or fertilising of the land, wheat was grown so cheaply that, even after the cost of transporting it over half a continent and the Atlantic Ocean, it could still be sold profitably in English markets at a price below that which British farmers were willing to accept. After the Repeal of the Corn Laws in 1845 the price of British bread depended more and more on good and bad harvests in the prairies of Canada and the United States, and in the Black Earth region of South Russia, and less and less on the vagaries of English weather. Thus in the later decades of the nineteenth century more and more old corn land at home was laid down to permanent grass.

In the literature of the middle nineteenth century a clear distinction is drawn between the corn counties and the grazing counties. The eastern counties of England, the East Riding, Lincoln, Cambridge, Norfolk, Suffolk, and Essex lying along and to the east of the sweep of the Yorkshire and Lincolnshire Wolds and the East Anglian Heights, kept more than half their acreage in arable: these together with the eastern lowlands of Scotland were the corn counties. The eastern and southern midland counties between the Chilterns and the Cotswolds belonged wholly neither to the corn counties nor to the grazing counties, but their agricultural pattern matched their soil pattern, arable predominated on the light, permanent grass on the heavier soils. As the century passed these counties tended to join the grazing counties. The rest of England and Scotland was already by 1850 recognised as a predominantly pastoral land and become more and more so in later decades.

The grazing counties were not uniform in their economy, considered either geographically or historically. Sheep were dominant in some areas, cattle in others. In the early and middle decades of the nineteenth century wool, meat and dairy produce were the main staples of Britain's pastoral economy; in the later decades pedigree stock and milk became increasingly important.

In the first half of the nineteenth century Britain was dependent on her own acres for most of the wool and nearly all the meat and dairy produce that she needed. Sheep farming was carried on on three different types of land: hill, downland and lowland. Hill sheep farming was of great importance in the Southern Uplands and the Border country; here Blackface on the heather, Cheviots on the grass hills managed to thrive on poor pasture producing fine sweet mutton and a wool, rather coarse but rich in oil, prized for its hard-wearing and waterproof qualities. Blackface flocks in the Scottish Highlands, Mountain Welsh flocks in Wales were increasing in number, encouraged by the big demand for wool; in both areas they were ousting the older cattle economy. Downland sheep were particularly numerous in Wiltshire and Dorset; there the Southdown sheep, the result of the earliest successful attempt to inbreed for a good quality, short-wooled fleece, were the favourites, valued for their fine fleecy wool and good quality mutton. Of Lowland sheep the most famous were the Leicesters, bred for their long wool to meet the demands of the worsted weavers of Norfolk and Yorkshire and fed on Leicester and Lincolnshire pastures. The Romney sheep, another long-wooled breed especially adapted to their peculiar environment, grazed the drained marshlands in Kent.

Cattle were kept in different districts primarily for different purposes: for stock in the most distant pastures, for dairying in the more accessible areas, to fatten for killing in pastures nearest the meat markets. The great stock rearing areas were in South-west Scotland, the Lake District, the Pennines, the Welsh borderland and Devon and Cornwall. From these relatively remote pastures droves of young cattle came south and east to supply the dairy farmers of Lancashire and Cheshire, of Staffordshire and Shropshire, of Somerset and Buckinghamshire, and the graziers of the eastern Midlands. On the fat pastures of Leicestershire, Northamptonshire, Rutland and Hertford beasts were fattened, then sold in prime condition to the butchers of the industrial cities and of London.

In the later nineteenth century the picture changes. As the Australian squatters established their sheep stations across New South Wales and Victoria, homegrown wool, already supplemented by merino wool from Saxony, became less important to Yorkshire buyers; with the introduction of refrigerated ships, beef from the Argentine, mutton from New Zealand reached the English market, and Canterbury came to mean "lamb", rather than an archbishopric or a piece of furniture, to thousands of British housewives. New Zealand and Danish butter soon rivalled in popularity the best farm produce locally churned. Cheese from distant pastures satisfied the palates of the many; "Cheshire", "Cheddar", "Stilton" and the rest were demanded only by the few. Even Empire and Danish eggs, Canadian and Danish bacon were soon to be found in plenty in British shops. British products were not wholly displaced; the home wool clip, it is true, soon became relatively unimportant, but home-grown cuts of beef and mutton could always command their price and there were no foreign substitutes for veal and lamb, nor incidentally for pork.

In spite of these changes, the Scottish and Welsh mountains remain important sheep farming districts; the mutton is sweet, lean and fine, the wool clip goes to make carpets and hard-wearing tweeds, and some of the Welsh wool, blankets and flannels; their pastures are too remote and too poor for cattle. Romney Marsh also retains its sheep, there is too little surface water for cattle. The Southdowns and Leicesters, however, lost their market for wool, Australian merino is finer than Southdown wool, and Australian long wool cheaper than the Leicester or Lincoln clip. The cattle raising districts remain little changed, but near the big towns and on pastures within quick reach of towns by railway there is a great increase in milk production. As communications improved fewer and fewer cowsheds remained within the cities and the unhappy-looking animals grazing on sooty town commons disappear. Milk production becomes so profitable that many districts take to dairying even where physical conditions are not wholly suitable. Drovers travelling south with their cattle saw the opportunity to grass unprofitable arable acres, to stock them with cows and to sell milk in town dairies. Many an Ayrshire farmer's son, for example, helped by his traditions chanced or ignored the effects of low rainfall and made plenty of money in the 'nineties dairy-farming in Essex, where Englishmen had failed to make ends meet on arable farms. In pastures too remote for the production of liquid milk, the quick way to make, or to lose, a fortune was to breed pedigree stock and export it to the new pastoral areas. Many a Galloway bull has pranced around Argentine pampas, many a Romney ewe bred in the Kentish marshes has later enjoyed salt licks on the Murrumbidgee.

Here and there, in favoured places, the old-fashioned brown of arable acres, the new lush green of good pasture land was replaced. The feathery green of carrots, the shiny green of onions, the bluish green of cabbages, and the silver sheen of glasshouses mark acres of market gardens around London and the big cities of the Midlands and the North. Apple, plum and cherry orchards add gaiety to Kent, the Vale of Evesham, and the Cambridge heights. The tulips of Lincolnshire, the daffodils of the Scilly Isles, however, bloom for the most part not on their native soil but in Covent Garden Market.

Enclosures

The third change, the enclosure of those parts of the land that in the eighteenth century still lay in common fields and common pastures, was closely related both as cause and effect to the new methods of husbandry. To hoe one set of strips carefully to keep them clean was heartbreaking work if neighbours left their strips weedy; new rotations could only be practised by all or none, for some to plant roots on their fallow strips was useless if others insisted on their rights of common and turned animals into a field to graze; to breed animals deliberately to foster certain traits was pointless so long as the carefully bred grazed horn under horn with any and every mongrel. High farming thus encouraged enclosure. On the other hand, it was essential to farm newly enclosed land well in order to get return for the money expended to pay for an Act to permit enclosure, to pay a surveyor to map the village lands and allot a share to each claimant, to pay for hedging or walling fields and perhaps for building a new farm house and steadings. Enclosure thus encouraged high farming.

The great period of change from open fields to enclosures lay between 1760 and 1830. Before 1760 some 270 Acts of Enclosure were passed, 156 of them in the last decade; between 1760 and 1810, 2,765 Acts of Enclosure public and private are to be found.

Enclosures did not, however, redress all England at this time. The greater part of the arable land to which the Enclosure Acts of the eighteenth and nineteenth century refer lies in a wide belt running across Midland England (Fig. 39). This area presumably until then lay, technically at least, in open fields. In addition the Acts refer in the North-West to large

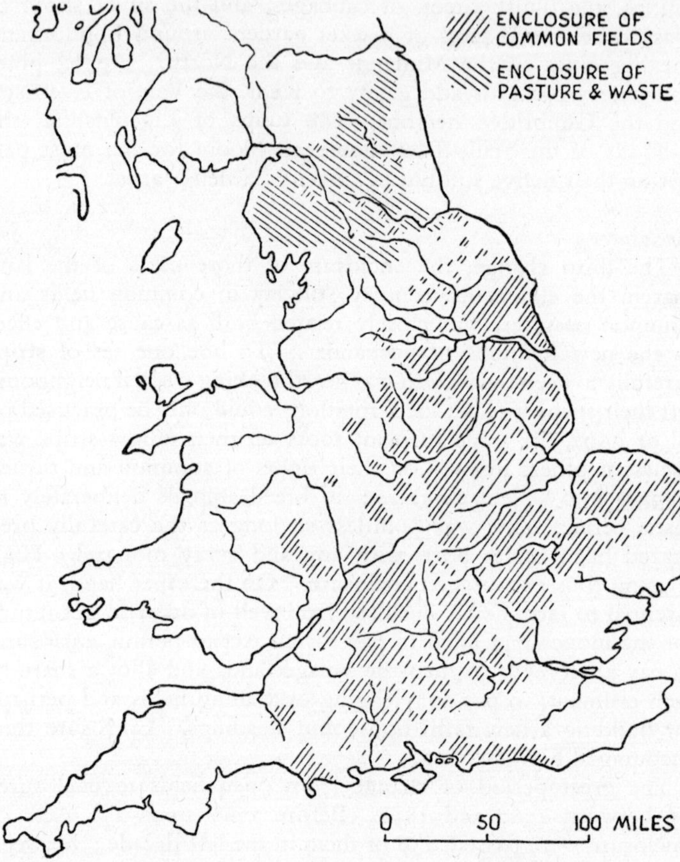

Fig. 39. Areas in which much Common Land was enclosed by Act.

tracts of mountain and hill land as yet little used except locally for rough grazing. In contrast, in the rest of the country the Acts refer only to small and scattered patches of common fields, commons and waste. Since there is very little land held in common among villagers today (with a very few exceptions) it follows that common of arable and pasture had largely disappeared in these areas at an earlier date. The enclosure of the land is then only the final phase of a movement which had progressed sometimes more, sometimes less rapidly ever since the settlement.

The regional distribution of open fields and enclosures at successive periods from the sixteenth century onwards can be established in very broad outline but there is no easy explanation of even the general features of it. In considering the problem to try to assess it, however tentatively, two things may help: the probable ease with which enclosures could be made on the one hand, and the probable stimulus to make the effort on the other.

The considerable areas of the north-west and south-east which have little or no mention in the Enclosure Acts may have been early enclosed. Old farm houses, numerous small fields, and roads that twist and turn every field length suggest in many districts of these areas early piecemeal enclosures. Tudor documentary evidence also suggests that enclosure in the north-west and south-east was at least not a problem in Tudor times. The "Act to Maintain Husbandry and Tillage" (39 Elizabeth C.2) 1597-8 omits from its provisions designed to prevent enclosure of arable for sheep pastures the counties of Cumberland, Westmorland, Lancashire, Cheshire, Staffordshire, Shropshire, Cornwall and Devon, and East Anglia, Kent, Surrey and Sussex. Francis Trigge in presenting his "Humble Petition" against enclosures specifically exempts from his charges "Essex, Hertfordshire, Devon and other Woodland counties where everie lordship is charitably divided among the tenants and tillage also in most of the closes is maintained and towns[1] nothing dispeopled". Later writers frequently refer to "old" and "ancient" enclosures in districts once well wooded. Scattered hamlets, divided parishes, few records for parliamentary enclosures seem to go together and all may be the result of secondary and late settlement. If old arable is already organised into three fields it can but rarely have been easy to extend cultivation by clearing an equal amount of land on the edge of each of the three

[1] "Towns" here is used to mean townships, parishes, not in the modern sense.

fields so that each can be enlarged. An outlying field newly cleared would usually have to be dealt with on its own and frequently no doubt it would be given to or taken by an individual as part or the whole of his share, rather than divided out between all the farmers of the village. The absence of enclosure acts concerning many parishes lying in the Weald of Kent, Surrey and Sussex, in the wooded parts of the Chiltern counties of Buckingham, Hertford and Essex, and in the woodlands of the Midlands in Cheshire, Shropshire, Worcestershire and Warwickshire may be thus explained.

In much of the north and west arable land was but an adjunct of the pastures and often lay in patches scattered throughout the parish. It must often have been convenient to acquire the right to the sole use of one arable patch and the surrounding pastures near the shieling built to minister to flocks and herds rather than to keep a share in all the arable fields. It could not have been difficult to agree about the allotment of the arable acres where people were few and wastes wide. In the process of division if more land was wanted to satisfy individual needs, and demands there was plenty for the taking.

In the good lands of the south and east the explanation of early enclosure may lie not in the ease with which enclosure could be effected but in the strength of the stimulus to make the change. It was here that wave after wave of settlers arrived, it was here that population early became relatively great, the stimulus to farm as intensively as knowledge allowed would be strong, for mere subsistence farming would demand a big return from the land. Then position favoured trade and a yet bigger return would allow farming for profit by the sale of surpluses in London and the nearby continental markets. The go-ahead would be quick to resent the restrictions of common field agriculture. It may be that by the time there is documentary evidence to tell how land was farmed the two- or three-field system, though suited to the natural environment of the south-east, no longer fitted the economic environment here and was already in decay. The field systems of many parts of Kent and East Anglia as revealed by manor records show well-marked peculiarities.

Elsewhere it was perhaps the very nicety of the balance

between man with a given technological equipment, the particular economic and social structure that he evolved, and the intrinsic conditions of the natural environment, that created stability. Where the settlers in each village early extended control over the land around until they came to land used by a neighbouring village, there boundaries were early fixed. Each group tilled the maximum amount of arable that could be cultivated by the number of beasts that could be pastured without destroying the modicum of wild needed for essential raw materials and game. There was little if any waste to come and go on. In these circumstances to divide common fields and pastures into individual holdings was very difficult. If the equivalent of each man's arable strips were allotted to him as a block, it might be sufficient for his need for ploughland. But there would then be no longer one large field in stubble and in fallow over which all the animals of the village could graze at stated times of the year. Animals, like men, can be fed more economically together in large numbers than in small separate groups, and thus the grazing on small patches of stubble and fallow for individual flocks and herds would not be an adequate substitute for the common of pasture. In many districts no extra pasture could be taken from the wild to make up. Here was the crux; the right to graze in common fields and on common pastures was the clamp that held the land in the two- or three-field system until the "new husbandry" with its roots and sown grasses for animal food brought release. The opposition to the creation of Royal Forests and the encroachment of the farmers around their edges and even within them and the agitations against uncontrolled assarting reveal that in some districts farming economy early reached a fine balance between the wild and the sown. The increasing demand for wool in the fourteenth and fifteenth centuries led some predominantly arable districts to attempt to increase their sheep flocks to share in the prosperity of the pasture lands. To increase flocks meant to decrease ploughlands and the clamour raised against enclosure for sheep farming was great: ballads were sung, sermons were preached, petitions carried to Parliament against the iniquities of the enclosures.

"Commons to close and kepe
Poor folk for bred to cry and wepe ;
Towns pulled down to pasture shepe ;
This is the new gyse."

ran the ballad "*Now-a-dayes*". One writer indicts particularly the "Middle Partes of the Realm" and a "Tract of Certain Causes of the Decay of England" written in 1550 cites enclosures, especially in Oxfordshire, Buckinghamshire and Northamptonshire. The system of two or three open arable fields was perhaps most firmly rooted in Midland England and thus farming methods were least adaptable here. Certainly numerous Acts of the Tudor period were passed against the conversion of arable to pasture and the legislation seems to have been on the whole successful in staying the changes here for it was in this area that most land lay open to be enclosed when policy changed in the eighteenth and nineteenth centuries.

It must not be thought that all the villages of the Midlands were open field villages, or even that all the farm land in any one village lay in open fields by the late eighteenth century. Where for one reason or another waste was available to give a certain elasticity to land use around the edges of fen, forest, moor or down, there villages or parts of villages are early enclosed. Where land lay near a growing town sometimes the stimulated demand for country produce brought high farming and enclosure, in spite of accompanying temporary and local disturbance. Where the land belongs for the most part to a few big landowners sufficiently strong to have their way, agreement to enclose is often reached earlier than where many holders must be satisfied. There are many local peculiarities in the way in which and in the time at which enclosure took place. This is clearly shown by F. G. Emmison, *Types of Open Field Systems in the Midlands* (Historical Association pamphlet), which illustrates the great variety of field arrangements seen in the Enclosure Award maps of Bedfordshire.

Open fields lingered longest in large areas of the Eastern Midlands ; in Northamptonshire and Huntingdonshire, in

Bedfordshire and Buckinghamshire, and perhaps longest of all in parts of Oxfordshire and Cambridgeshire. Even in 1822, entering Cambridgeshire at Royston, Cobbett could talk of " those very ugly things common fields that have all the nakedness without any of the smoothness of the Downs ".

Many more regional studies are needed before the general pattern of the distribution of the open fields and the progress of the enclosure movement is fully worked out. M. Aurousseau in an article on " A Neglected Aspect of the Enclosure Movement ", *Economic History*, Vol. I, 1928, emphasises the value of a geographical approach to enclosure problems. Detailed work on enclosures is perhaps rather difficult for a beginner, but in most regions some attempt to establish in general the distribution of open fields and enclosures in the eighteenth century is interesting and profitable and not too difficult. Local topographers usually comment on the enclosures and most of the county Reports of the Board of Agriculture discuss them. The descriptions of the local scene by Leland or Camden, by Defoe or Young or other travellers, may also help. Ogilby's road survey distinguishes by pecked lines the roads through open land, a practice followed by most later cartographers. Where enclosures are late the date at which the parishes are enclosed by Act may be plotted. The maps when finished will show whether the enclosures of relatively early and later dates are scattered apparently at random or whether they form a definite geographical pattern. The interpretation of the pattern is more difficult. The maps should be studied in conjunction with relief and soil maps, with population density maps, with maps showing roads and navigable water and thus the accessibility of the area, and with maps of land ownership. The correlations between these maps should be pondered, but it is necessary to be chary of assuming direct causal links to explain them. One thing must always be remembered; although the Enclosure Act and the enclosure movement are usually more or less contemporaneous, they may not be. Sometimes the Act of Enclosure merely confers a statutory blessing on a division of the village lands completed long ago; sometimes, though permission to enclose is granted by Act, the villagers may long delay availing themselves of it. To pursue the study further, the Enclosure Awards and the maps that accompany them if they survive must themselves be examined. In some areas Awards and maps are extant in numbers, in others there are very few. In individual villages one can often see the same general factors already considered operating, the ease of enclosure where there is waste, the stimulus to enclose where there is dense population.

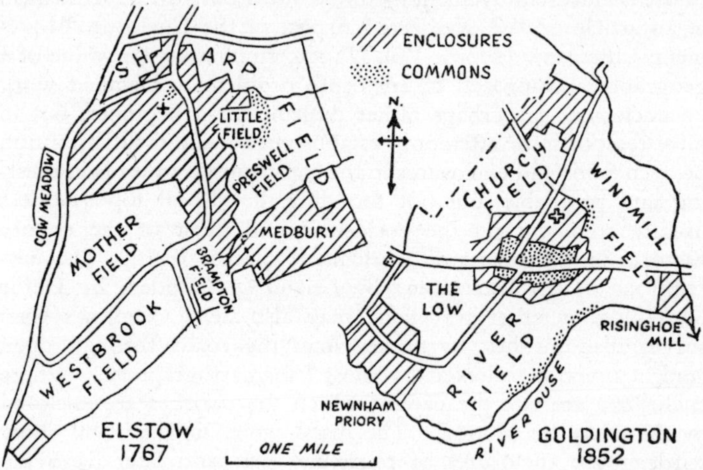

Fig. 40. Diagram of Pre-Enclosure Award maps of Elston and Goldington, Bedfordshire (from F. G. Emmison, " Types of Open Field Parishes in the Midlands", 1937.)

Fields early enclosed are frequently found in two groups: on the edge of the parish where late assarting from the waste took place, and around the village where closes for animals and orchards and gardens were wanted (Fig. 40).

The clearing of the woodland, the draining of the marshland, and the transformation of the cultivated landscape as new techniques are evolved to meet new needs, these all-important phases of historical geography can be similarly studied in any other region if the ground is known, the language understood, and the documents accessible. In Mediterranean lands the transformation of the scene by permanently settled human groups began so long ago that the details of the process are lost now. Neolithic farming peoples settled around Mediterranean shores in the sixth millennia B.C., occupying first the "middle slopes", above the coastal and river marshes where waterlogged soils inhibited the cultivation of both grain and fruit trees, and below the steep forested upper slopes where mountain soils were thin, and if cleared were soon eroded. By the time of the classical writers the Mediterranean landscape had already acquired many of its present characteristic features; man with primitive ploughs cultivating grain, with spade and mattock cultivating vines and olives, with sheep and goat grazing marsh and mountain pastures, had destroyed most of the natural vegetation, had made already the bare hillsides patched by fields. Highland divides separated village from village; these alone were left untouched, and in parts of Greece and Spain are little changed to this day.

In Mediterranean lands changes in agricultural technique have not been great since classical times. The Arabs were at least as important as renovators as innovators, though Mediterranean farmers may owe to them some of the tropical crops, for example rice, cotton and sugar cane. To watch the work of an Andalusian cultivator, a Sicilian farmer, a Greek gardener or Albanian shepherd today is often to see the living practice of the precepts of classical texts. "New husbandry" and "high farming" did not here transform the rural scene in the eighteenth or nineteenth century, not altogether because of poverty. In many ways "high farming" came here long ago; a nice balance was early achieved between man and nature; to alter the way of life might well have been to alter for the worse. Animal and plant breeding, scientific manuring, and the mechanisation of farming is only now in the twentieth century bringing "improvement" slowly to this area.

The age of full settlement north and west of the Pyrenees, Alps and Carpathians is still great, though less than about the Mediterranean. The Neolithic farmers colonising the most easily cultivated soils of the Danube valley, of the northern foot-hills of the Carpathian and Bohemian mountains below the forest and above the marsh, appear to have reached the loamy soils of the river terraces of the Rhineland, of the Schelt-Lys basin, and of the Seine somewhere between 3000 and 2000 B.C. Celts colonising north and west from the Swiss lake region, Teutons settling in the west, Slavs in the east, had by the dawn of historic times occupied the most attractive patches of light and medium soil, and during the early centuries of our era settled the more accessible and fertile parts of the clay lands. The later progress of colonisation in France and western Germany and the Low Countries was here, as in England, from these early settled areas into the forests of the heavy clay soils, the woods of the mountain slopes, and the marshes of the river flats and coastal belts. The general trend has been continuous gain of arable and pasture land at the expense of wood, moor and marsh; but in detail the frontier between sown and waste has shown advances and retreats. Periods of peace show advance, periods of war and pestilence show retreat; the fluctuations have been perhaps on a larger scale in Europe than in England as times were more turbulent. One of the great, perhaps the greatest, period of assarting and draining in north-western Europe falls within the three hundred years 1050-1350. In the Low Countries as in the English Fenland a complete transformation of the land was brought about by marsh reclamation. The land was colonised slowly and painfully in a long and constant struggle against invading sea, flooding rivers, and waterlogged soil. Ambitious and extensive schemes were carried out much earlier here than in England, much land was drained by the end of the ninth century. Dutch schemes provided the model and Dutch engineers the experience used not only elsewhere in Europe but in England and the New World too. The Dutch on their hard-won but fertile lands were progressive farmers and many of the new crops and new practices that characterised the "new husbandry" in England had their origin in Holland. In the upper valleys of the tributaries of the Danube and in the North German plain east of the Elbe, clearance of woodland and reclamation of moor and marsh has gone on from prehistoric to modern times. The higher parts of the Hercynian mountain blocks remain forest land, but the "limitless" Hercynian forest described by Tacitus was largely cleared in the twelfth and thirteenth centuries. The "*Waldhufendorf*", as the Germans call the forest villages the medieval colonists planted, are still part of the landscape of the lower slopes of the Alps, the Jura and the Böhmerwald, of the forest lands of Thuringia, the Erz Gebirge and the Sudetenland, below the zone of the hamlet settlements of the upper pastures. The Flemings were urged to come and drain the marshes; villages of Flemish character, often with Flemish names, are found around the mouths of the Weser, Elbe and Oder, about the river confluence of Elbe and Saale, and in the marshes of the Havel and the Spree; these are of medieval date. Projects to drain large areas of coastal marsh in the Lower Vistula and in the Warthe-Netze valley also date from the thirteenth century but they were not carried out until the eighteenth or nineteenth century; the pressure of population upon the land within the area and the attraction of it to colonists from without was not great enough.

In Europe the slow transformation of the countryside by the work of man can be tentatively traced by piecing together varied and scattered evidence. But in "new" countries changes fully documented take place within decades that it has taken all the centuries of recorded history to bring about in the old lands. The frontier of European settlement in the United States in 1790 was still close to the Atlantic seaboard, by 1830 it had reached the Mississippi valley, by 1860 the edge of the Great Plains, and by 1890 the frontier, in the sense of a continuous line that can be drawn through country with a density of population of less than two people per square mile, had disappeared. With the advance of the frontier came not only the clearance of the forests and the breaking up of the prairie sod, but a constantly changing economy too. There is plenty of literature to allow us to follow the hunter and trapper as he pitches his tent, the pioneer farmer as he builds his log cabin, the surveyor as he lays out his township; each pushes the frontier a step further west, each pushes the settlement a stage further on. And behind them, into the "fallen down" areas drift some of the least well-endowed immigrants to wring some kind of meagre living from abandoned fields, and often to convert the merely exhausted and overworked soil into eroded and barren land. But into these badly used farms too, near cities or within easy reach of good markets, have come lately the economical, hard working Scandinavian farmer and the skilled Italian market gardener, who with science and labour have made three carrots grow where not one blade of grass grew before. Truck farming on a luxury scale is now another frontier of immigrant expansion.

The changing face of the English countryside, it has been emphasised, can only be studied intelligently in relation to man's work in Europe, in America, in the Antipodes and the Tropics. On the other hand, and just as important, the geographical changes wrought by man, be he settler, governor or missionary, reflect not only his response to his new natural environment but also the conditions of his old environment. The geography of the " new " lands has a far longer history than the episodes that have occurred within the area. The tradition of the long straight furrow that ran up and down the hill to give good drainage in the cool temperate climate of western Europe, transplanted to Virginia and Carolina, caused much of the soil erosion that succeeded soil exhaustion when worn-out tobacco lands were ploughed and sown with grain. The strong belief bred in Europe in the infertility of land that would not grow trees held the settlers back on the edge of the prairie soils as firmly as the lack of suitable ploughs or building materials. The tradition of " constant cultivation " inculcated by the best farmers of eighteenth-century England is much to blame for the creation of the Dust Bowl in the States of the Great Plains, and the devotion to bare fallow as a rest cure for land has undoubtedly contributed to the advance of the desert.

In tropical lands, Europeans found a still more unfamiliar environment. The traditions and experience gained at home were there often a positive disadvantage, especially if settlers strongly upheld and stoutly defended the introduction of practices they considered " progressive " in the face of resistance from natives. The indigenous groups, though without appreciation of " scientific " agriculture and unmoved by European economic needs, had often a better understanding from their own experience of how to co-operate successfully with Nature in the tropics. Increasing biological knowledge in general, and in particular the accumulation of facts concerning food chains and their relation to optimum plant, animal and human populations, are bringing about a better grasp of the close and complex inter-actions and re-actions that occur if any one element of the environment is altered. The necessity to hurry slowly, the realisation that he can change and alter safely only after careful thought and full experiment if unforeseen and unwanted side effects are not to nullify the advantage expected, are being forcibly borne in upon man, especially in environments with which he is not wholly familiar. He now possesses techniques so potent in inducing far-reaching changes that they cannot be used lightly; the study of the historical geography of alien environments that he has already colonised, should provide due warning and teach wisdom through the experience of earlier generations.

Map Exercises

Suggestions of maps the reader may like to make for his own area where appropriate.

1. Map of wooded areas on present O.S. sheet, dating where possible the planting of the woods.
2. Map of wooded areas on 1st edition O.S. sheet.
3. Map of wooded areas shown by Saxton or Speed.
4. Distribution of vills with mention of wood in Domesday Survey.
5. Distribution of place-names containing elements referring to woodland, clearings, or trees.
6. Distribution of place-names containing elements referring to marshland.
7. Map of artificial waterways cut to drain the land.
8. Map of dates of reclamation of moorland, heath or down.
9. Map of dates of Enclosure Awards referring to parishes of district.
10. Map of "closes" and open fields of selected individual parishes from Enclosure Award Maps.

CHAPTER VII

THE CHANGING GEOGRAPHY OF INDUSTRY

THE location of industry depends very largely upon the abundance of the raw materials found or produced within, or easily imported to, the area; the physical conditions required by the techniques of the time; the amount and skill of local labour; and the position of the markets. The advantages of position in relation to the source of raw materials and the markets for finished goods are apt to alter with changes in supply and demand and changes in conditions of transport. The evaluation of the suitability of the site is apt to alter with changes in methods of manufacture. Therefore the geographical pattern of industry shows considerable and often rapid variations from time to time and gives an opportunity, greater in the industrial than the agricultural scene, to examine the geographical effects of economic and technical change. On the other hand a successful industry creates within an area conditions of capital and labour that tend towards stability; an industry once established has strong roots. The attributes of site and situation which favoured its establishment and early growth may become of no importance or even may become positively disadvantageous, and yet the investment of capital and the economic and social organisation built up may serve to tie the industry securely to its birthplace. Even if the original industry declines and eventually dies out, similar or other industries often establish themselves in its place using the slack of labour and capital that the declining one leaves. The manufacture of motor cars in Coventry, for instance, is but the latest of many successive different industries in that city. The historical legacy of industrialisation is thus in itself an extremely strong factor in considering the location of modern industry, and this again demands the approach of historical geography.

Local Study : Sources

Regions vary too much in their industrial pattern for it to be possible to give much useful general advice. In areas where industry has grown up in the eighteenth and nineteenth centuries sources for the study of its geography in successive decades are usually abundant and do not differ very greatly in kind from the sources used in making a similar study of present day conditions. There are a great many good modern books and articles dealing with the recent history of particular regions or particular industries, and much can be learnt too from the descriptive, technical and biographical writings of the period under investigation. Census returns and other Government statistical publications, and reports of Parliamentary Committees and Royal Commissions are a mine of information. The ground itself, the factory buildings and the houses of the workers and owners, by their date, size and distribution give further data. In these areas, changes since the Industrial Revolution have been so great that they obliterate almost entirely earlier patterns, whether of an agricultural or an industrial scene. In other areas where industry has been a characteristic of the land in earlier times but in which agriculture has remained or become re-established as the mainstay of life, sources of direct information may be much less abundant; many and varied clues must be followed up to establish the industrial geography of earlier days and to appreciate the legacy it has left. Surviving taxation lists—national and local, parish registers, church and domestic architecture will usually provide some information to supplement the story told by local historians or by curious passing travellers of the day.

The Historical Geography of the Textile Industries and the Iron Industry in Britain

The changing geography of the textile industries and of the iron and steel industries allows illustration of the importance in studying the problem of location of considering the economic and social needs and opportunities of the time in conjunction with the physical conditions of the place. The time must be ripe as well as the place suitable if a new industry or industrial technique is to take root. The cotton industry is used in particular to illustrate this point. The geography of the textile and iron and steel industries affords also plentiful illustration of the legacies left to modern times by earlier industrial patterns, and of the geographical revaluations brought about by technological advances. The geography of the woollen and worsted industry before the Industrial Revolution is discussed to exemplify in particular the first point, and the rapid changes in the geography of the iron and steel industry the second.

The Textile Industries

The textile industries show today a well-defined geographical pattern; the cotton industry is strongly localised in South-East Lancashire, the woollen industry in the West Riding of Yorkshire. Silk and the rayon and newer synthetic fibres are spun and woven in many scattered factories with some concentration, to some extent as replacement industries, in and around the Lancashire cotton towns.

The Cotton Industry

The cotton industry shows today a well-marked geographical division between its two main branches, spinning and weaving: spinning is for the most part carried on in the horseshoe of towns around Manchester, from Wigan through Bolton, Bury, Rochdale, Oldham to Stalybridge; weaving is dominantly the occupation of the region north of Rossendale and south of the Ribble, of the towns of Preston, Blackburn, Accrington and Burnley. Some understanding of the location of the cotton industry in general, and of its two processes in particular, can only be achieved if consideration is given to the evaluation of the opportunities of the place in relation to the conditions of the time. It is not sufficient to cite the damp climate of Lancashire with its high humidity on most days throughout the year, and to note that on rainfall maps the precipitation is higher to the west than to the east of the Pennines, to explain the dominance in the west of cotton manufacture with its finer threads and less natural grease to aid cohesion of the fibres. It is not sufficient to plot the index figures to show hardness of water and to note that the cotton towns lie to the south of the Ribble where the water is soft. It is not enough to plot the coal fields and to show that the cotton towns lie on or near them. These facts may play a part in determining the distribution of the cotton towns but they do not primarily explain it. Factories nowadays can artificially make the atmosphere in which thread is spun of optimum humidity. The temperature and humidity of the air around the spinning machines may bear no relationship to that of the air in the world outside, or even to that in other parts of the factory. Once, everybody and everything had to make do with the air as Nature provided it; now, at some cost it is true, air can be altered to suit the thread and, if the office staff does not flourish in conditions optimum for spinning, their air can be different and as they like it. Soft water, if it does not exist within the area, can be piped to it from another area or the local hard water can be artificially softened to any degree required. Power, whether in the form of coal to generate steam on the spot or in the form of electricity can be brought to the factory; the factory is no longer tied geographically to the coal seam or the river current. In so far as natural physical conditions have an effect on present industry, it is largely through the effect they exerted on past industry; they operate now through the strength of geographical inertia arising from historical momentum. However, these physical conditions alone will not suffice to explain the location of the industry present or past; to attempt to make them do so is to over-simplify the links between work and place and to distort the relationship. The air in West Yorkshire was sufficiently humid to spin cotton successfully, water was soft there too, and coal lay accessible under the surface. The differentiation between the cotton and woollen manufacturing areas must therefore depend upon something else.

To examine the historical geography of the cotton industry does throw some light on the problem of its location. The time at which three innovations which affected it came is of special significance: the time of the introduction of cotton fibre as an important new material in textile manufacture; the time of the invention of simple water-driven spinning and weaving machines; and the time of the change from water power to steam power. The introduction of new materials, of new machines, of new sources of power will only be of significance at any one time to those areas that have people ready to make use of the innovations. Where people are too remote to learn about them, too backward to realise their value, too poor to face the initial cost of installation, too busy with profitable ploys of their own to be interested in new possibilities, there no use will be made of the inventions.

The cotton industry finds a home in Lancashire essentially because here, at the time of the *introduction* of cotton fibres, Lancashire was ripe for the development of a new industry. Agriculture alone was not a wholly satisfactory economy here; the climate and soil did not favour bumper wheat or barley crops as in the "corn" counties. Lancashire pastures and Lancashire sheep did not earn big revenues from wool as in the Welsh border lands, the Cotswolds and Leicestershire. Its traditional industries were not wholly satisfactory either; its woollens were not as good as those of Yorkshire and East Anglia, let alone of the West country. They suffered perhaps from the relatively poor quality of the local wool and also from the remoteness of the district, which starved it both of some essential raw materials of first quality—for example oil, dyestuffs and fuller's earth—and of the stimulus of ready markets supplied by a dense local population and by Continental links. Manchester "cottons" and friezes were coarse light-weight stuffs with a raised nap (to cotton or to frieze means to raise a nap). They supplied a local market and there was some trade in them with the less fastidious foreign markets unable to afford fine stuffs. The Continental wars of the late sixteenth century and of the seventeenth century upset the foreign market and brought distress to this as to other areas, and Lancashire like East Anglia was ready to try new cloths.

Cotton fibres in the sense of "being a fruit of the earth growing upon little bushes or shrubs" were not at first used in Europe to make a pure cotton cloth, but "Tree wool", as the Germans called it, was used as the weft on either a wool, linen or silk warp to make "fustians". By the twelfth century fustians were made in Italy using cotton imported from Sicily and the Levant;

fustians were early made in Spain too and both countries exported to England. Chaucer's knight wears a " gepoun of fustyan ". In the fourteenth century, with the growth of trade across the Alps, the industry became important in Switzerland and in south Germany, particularly at Ulm and Augsburg. England imported fustians largely from Genoa and Ulm, hence the common names " Jeans " and " Holmes " for these cloths. The manufacture was taken up in the Low Countries and it may be that the Flemish and Dutch refugees brought the technique of making fustians to England along with many other " new draperies ". In 1620 there are records of the export of " ordinary English fustians " and of " Milan fustians of English making ". Lancashire and Norfolk both had medieval linen industries as well as woollen industries and perhaps already had a tradition of making mixed fabrics, which served as a base on which to graft this particular "new drapery". In Norfolk, however, fustian never became of first importance; good quality wool could readily be bought in Leicestershire and Lincolnshire; it was long in staple and allowed the making of fine yarn. A silk weft mixes well with fine woollen warp, and silk and wool fabrics became the Norfolk speciality. But in Lancashire, in a belt of country along the border of the area in which linen manufacture early predominated in the west and woollen manufactures on the east, the fustian industry flourished, true woollen manufactures retreating into the upper valleys. The demand for fustians was increased by the prohibition in 1720 of the import of printed calicoes and muslins from the Far East, a prohibition made in the interests of the woollen and silk manufacturers but which encouraged home manufacture of cottons too.

The fustian industry was at first restricted by the limited amount of cotton it paid to import from Cyprus and Smyrna to London, and thence to Lancashire by packhorse. A new source of raw material helped Lancashire. Already in the seventeenth century there was some import from Brazil, and after 1740 the long staple sea island cotton was grown in South Carolina, and later in Georgia. The upland cotton which would grow well in all lowland areas of southern U.S.A. was short in staple and difficult to " clean ", i.e. to separate the cotton fibres from the seed pod.

Then came a mechanical invention: Eli Whitney in 1791 made his gin. A good slave could clean by hand 5 lb. of cotton a day; with the gin he could clean 50 lb. and once the gin was operated by water-power 1,000 lb. Cotton planting spread rapidly, and greatly encouraged the settlement of Alabama and Louisiana, and the annexation of Texas. Raw cotton could be produced in quantity at a price Lancashire could afford to pay, and was produced in a place with which Lancashire's growing port, Liverpool, was already closely in touch.

Great quantities of cotton would however not have been wanted if the thread spun from it could only have been made by hand and used as the weft in mixed fabrics. Mechanisation changed the medieval into the modern industry. In 1767 Hargreaves invented his spinning "jenny" (called for his daughter); this allowed one worker to attend to a number of spindles instead of one. The use of spinning jennies immensely increased output of yarn, but of a yarn not strong enough for the warp. Arkwright in 1769 brought into use a spinning machine that spun threads by means of rollers revolving at different rates. The machine could be worked by water power and became known as the water frame. This could spin strong if rather coarse yarn suitable for the warp as well as the weft. Crompton in 1779 invented his "mule" which combined the principles of the jenny and the water frame; it could spin strong but fine yarns, was worked by a water wheel and was not patented, so became immensely popular. Although Hargreaves, Arkwright and Crompton were Lancashire men, their inventions were first taken up with enthusiasm by the Nottingham hosiers. When they had proved a success in spinning knitting yarn, their use spread among the cotton spinners. Mechanisation meant factories; the machines except the very simplest were too costly to be owned by individual spinners. Spinning factories were established in the old fustian-making area around Manchester; and, as Arkwright's water frame and Crompton's mule were used, the factories were established on streams, strung out along the valleys rather than grouped in towns. Manchester, therefore, became the centre of trade in raw cotton and in yarns rather than the centre of production of the yarn.

Weaving was still a cottage industry widely scattered among the farming population. A simple hand loom to weave the cloth, soft water in which to wash the cloth, clean hill sides on which to bleach the cloth in sun and wind, sometimes after soaking or boiling it in sour milk, were the only requirements, until about 1787 Cartwright invented his mechanical loom. Machine looms, however, came into use very slowly; weaving was a highly skilled operation not easily done by machine, and for long the hand weavers could keep pace with the output of machine spun yarn. When machine looms did come into use weaving too became a factory industry, and in Lancashire again a coincidence of place and time brought about a marked localisation. The country around Manchester was already engaged in spinning and it, earlier than the north, tried out power looms on the coarser cloths but wages were high; the area to the north of Rossendale had spare agricultural labour ready to take up industry, a tradition of hand loom weaving, and a reputation for fine fabrics. It hesitated to set up spinning factories in competition with going concerns in the south. The desire for factory work coincided with the success of weaving machines, so the new industry found favour here, and weaving became relatively more important than spinning in the northern part of the cloth-making area simply because there there was space for it. The early weaving mills, unlike the early spinning mills, were not localised in detail by the need for water power since by the time they came into general use they were operated by steam power.

The steam engine, the railways, and new chemical processes strengthened in general though they modified in detail the pattern of industry in Lancashire. With the invention of the steam engine, coal could be used to heat water to generate steam; the rate of flow of the water and the shape of the valley mattered less, the softness of the water mattered more. Hard water furred up the pipes and the boilers. New factories were placed on the coalfields where mines could easily be worked and water was soft; fortunately water flowing through most coal measure rocks is soft. Old factories to which coal could easily be brought by canal and later by railway remained in being. The building of canals and railways in Lancashire was at first largely a response to the need to move coal ; the gradients in the south-east were not too steep for these new highways. The pattern of industry in both weaving and spinning districts loosened out a little ; factories were no longer crowded into the upper and less accessible parts of the valleys. Advances in chemistry transformed the bleaching processes, and again South-West Lancashire could take advantage of the improvements. Salt could be brought by canal from Cheshire mines, lime from Derbyshire, and sulphur could be imported via Liverpool and sent upstream to Widnes and Runcorn. A flourishing chemical industry with an assured market in the cotton-bleaching industry grew rapidly. Cottons could now be bleached in a few days indoors instead of requiring exposure for months to wind and rain and sun on the hillsides. The new process fitted in. To bleach in the old way near factories belching smoke was rapidly becoming impossible. Dyeing and colour printing became closely linked with the chemical industry too. The great diversity of cloths produced kept the industry specialised. Each cloth needed different kinds of threads—fine counts, medium counts, coarse counts—and different treatment in finishing processes. Some firms therefore specialised in spinning, others in weaving special fabrics, others in bleaching, others in dyeing, others in printing, others in calendering. The details of the distribution of the different branches of the industry make a complex and fascinating study that must be looked at in relation first to the physical conditions of the place, secondly to the technological equipment of the masters and men for every stage from planting seed to selling cloth, and thirdly to the changing demands of the market, some in response to constant trends in population growth and increasing standards of living, others the result of Fashion's fickle tastes.

The rapid growth of the cotton industry coincided with a rapid growth of population in England: in 1700 it is calculated that the population of England was about 6½ million, in 1801 it was counted as 8½ million, in 1831 as 13.8 million. Standards of hygiene and standards of living were rising : more people wore underclothes and nightclothes, soap was becoming cheaper and more people washed their clothes regularly. Cottons washed better than woollens and the demand for them grew. The East India Company, prohibited from importing printed calicoes made in India to Britain, sold them where they could on the way home, thus encouraging a market for them in Africa. When Company rule ended in India conditions changed; and the market created for Indian cottons in Africa fell into Lancashire hands. Cool garments were needed; the missionaries associated nakedness with paganism and sin, the merchants encouraged the missionaries and clothes spread with Christianity and so-called "civilisation". Even in India and the Far East machine-produced Lancashire cottons competed with the silks and with the local hand-made cottons—the fine calicoes and muslins.

The strong localisation of the cotton industry in Lancashire is thus the result of a coincidence of local needs and local conditions, themselves the outcome of place and time, with world-wide needs and conditions. The more complex the interrelationship built up, the more difficult is radical change: the pattern in itself once established creates in turn conditions fashioned to meet its needs. When the technique of manufacturing a cloth made in part of cotton was brought to England, Lancashire had no great speciality of her own, but she possessed a tradition of working in mixed fabrics. In the eighteenth century when America took to cotton planting Lancashire possessed a port already engaged in Atlantic trade and in the nineteenth century, when mechanisation became the order of the day, Lancashire had a young industry willing to experiment. By no means all the factors which have fashioned the pattern of the cotton industry have been considered, but sufficient has been said to show that an understanding of the industrial geography of southern Lancashire involves consideration of Lancashire farmers as well as Lancashire spinners and weavers, merchants and sailors; of farmers and merchants in the Levant and manufacturers in Southern and Central Europe; of slave traders in West Africa and cotton planters in America; of villagers in India, China and Japan; of the changing fortunes of all these as the centuries pass. Lancashire's industrial towns are thus linked to the

" Cotton Lands " of Louisiana and Texas, to African forests, and to the troubled world of the Far East. They are linked too to the fortunes of woollen and worsted manufacturers in South-West England, East Anglia and Yorkshire.

The Woollen Industry

The concentration of the woollen and worsted industry today in south-west Yorkshire, effectively in the Aire and Calder valleys, is almost as strong as the concentration of the cotton industry in south-east Lancashire. The West of England suitings, the Welsh and Cumberland cloths, the tweeds of Cheviot and Harris—in so far as they are made locally and are not mere trade names for particular kinds of cloth made too in Yorkshire—have today a very restricted market. This concentration, like that of the cotton industry, has no simple explanation. Yorkshire does not owe her success in competition with other regions important in cloth manufacture in the Middle Ages to the fact that coal lay near at hand. The Gloucestershire coalfield lay almost as near to the Stroud valley as the Yorkshire mines to the Aire and Calder. It might have been a little expensive but not prohibitively so to import coal by water to Norwich from Newcastle. Yorkshire was drawing ahead before the use of steam power linked the textile industry to the coalfields. If any one factor is singled out as operative in giving the lead in cloth production to the West Riding, it is probable that it should be the willingness to use machines driven by the water-wheel. To state this however but leads to consideration of the reasons which allowed or stimulated Yorkshiremen thus early to mechanise their industry.

Yorkshire had a long tradition of cloth making: in the Middle Ages she made " cloth " of great variety but she made little " worsted ", and even when the " new draperies " were introduced from the Low Countries in the sixteenth century, Yorkshire at first paid little attention to them. By the eighteenth century, Yorkshire was manufacturing " new draperies " or " worsteds " too, though of less good quality than those of Norfolk. There is an essential distinction going back to early times between the two kinds of woollen fabrics, a distinction determined by the way in which the yarn was prepared for spinning. Wool before spinning was either "carded", i.e. the fibres were rubbed together between two boards set with bristles to make a rove, or "combed", i.e. straightened with a comb to lay the fibres parallel ready to be twisted by the spinner. Cloth made from yarn spun from carded wool was thickened and felted by fulling, and then stretched into shape, teasled to raise a nap, sheered to give an even surface to the pile, and finally pressed. Thus was made all the "cloth" or "woollens" of the documents. Fabric made of yarn spun from combed wool was not fulled, teased or sheared, but merely washed and pressed; it was never referred to as "cloth" or "woollens", but as "worsted" or "stuff". The "new draperies" were also made from combed wool.

Yorkshire perhaps profited by proximity to Lancashire; Yorkshiremen learnt of Lancashire's machines, and they had in their worsteds a cloth suitable for experiment in new methods of manufacture. Worsted yarn, made from combed wool, was more like cotton yarn than woollen yarn—it was smoother and more regular and easier for a machine to handle. The Yorkshire worsteds had as yet no great reputation, so the manufacturers could afford to attempt to cut their costs still further by mechanisation and had little to lose if early experiments went wrong. Early experiments did not, by and large, go wrong and mechanisation spread from the worsted industry to the woollen industry, and ultimately the mechanically spun and woven worsteds and woollens of Yorkshire equalled and even surpassed in quality the Norfolk stuffs and West Country cloths and could be produced at a fraction of the price. Yorkshire therefore captured the bulk of the trade of both regions. The transference from water power to steam power did not bring about the strong localisation of the woollen industries in Yorkshire but allowed, as in Lancashire, a greater dispersal of the factories within the region. Industry which had been creeping higher and higher up the valleys in search of a greater fall of water for power, on the introduction of steam power spread downstream again to more accessible and flatter sites.

The Geographical Legacy of the earlier distribution patterns of the Woollen Industry

When Yorkshire became pre-eminent in all branches of the woollen industry, all traces of earlier patterns of distribution were not wiped out. Earlier industry not only affected the growth and structure of many towns and villages but its imprint is to be seen too on the fields and pastures of the land. The wealth of Norfolk and Norwich is now fundamentally in its agriculture; it may always have been so. But industry in medieval and early modern times added considerably to the prosperity of the county and its legacy is clear. Some of the present industries of Norwich arise from processing agricultural products of the neighbourhood, the manufacture of mustard, starch and beer, for example. The manufacture of silk crapes on the other hand is a survival from the earlier period, and perhaps the wire drawing, and even the boot and shoe manufactures may have indirect links with the weaving of fine stuffs through the tradition of careful skills which, left unused by the decline of the old industries, could be used by the new. The towns of south-east Norfolk mentioned by Defoe (see p. 250) are certainly essentially country market towns, but country market towns which owe something of their character to their industrial importance in the eighteenth century. The villages of the middle Bure and its tributaries which were the centres of the flourishing medieval industry, still stand out from their neighbours if only by virtue of the splendour and size of their churches. Worstead gave its name for all time to those stuffs woven from combed wool for which the county was famous. In Suffolk and Essex, in the valleys of the Colne and the Stour, medieval industry has also left its mark on almost every town and village from Ipswich and Colchester to Lavenham and Clare, Thaxted and Saffron Walden. The weaving of horsehair and coconut matting at Lavenham and Long Melford, the making of scrims and dusters at Haverhill and the silk industry at Sudbury are in the direct tradition. In the West Country, the valleys of the South Cotswolds, of the Mendips, of the Marlborough Downs and the Salisbury Plain, of the Parrett and the Exe were once important industrial regions and their present character bears evident witness to this fact.

The changing geography of the woollen industry before the Industrial Revolution may therefore be of relevance in considering the present geography of many areas other than Yorkshire. Much of the cloth made in early and medieval Britain was not produced for market at all. In the British climate everyone needed clothes, early Britons clad in woad are figures of the imagination; sheepskin would have been more comfortable but uneconomical, thus the making of cloth from the wool or hair of animals or the fibres of plants is an early art. When " Adam delved " he also wove, and certainly " Eve span ". Every unmarried woman is still legally "a spinster". Perhaps every household once made its own cloth, as isolated households did for long, but the common surnames, Weaver, Webster, Fuller, Tucker, Walker, Dyer, Tozer, Shearer, Shearman, and the rest must belong to a time when there was some specialisation. The village " weaver " became as individual a figure as the village " smith ". Even before the Norman Conquest cloth may have been made for the export market. There is the famous letter from Charlemagne to Offa, King of Mercia, requesting that cloaks should be sent " of the same pattern that used to come to us of old time ". It is not, however, until the twelfth century that there is sufficient evidence of *where* cloth was made in more than usual quantity. Thereafter three periods with different patterns of distribution may be distinguished: in the eleventh and twelfth centuries certain towns would appear to lead in cloth manufacture; in the later Middle Ages and the Tudor period the industry was widely distributed in country districts; and in the seventeenth and eighteenth century there was a more strongly localised country industry.

In the twelfth century London, Winchester, Oxford, Huntingdon, Nottingham and Lincoln had guilds of weavers. Winchester had a guild of fullers too, and " laws of weavers and fullers " exist for Winchester, Marlborough, Beverley and York. Some of these towns were famous for their cloth: Beverley blues and Lincoln scarlets appear in Royal accounts of the twelfth century and in 1265 English Stamfords, dyed Stamfords

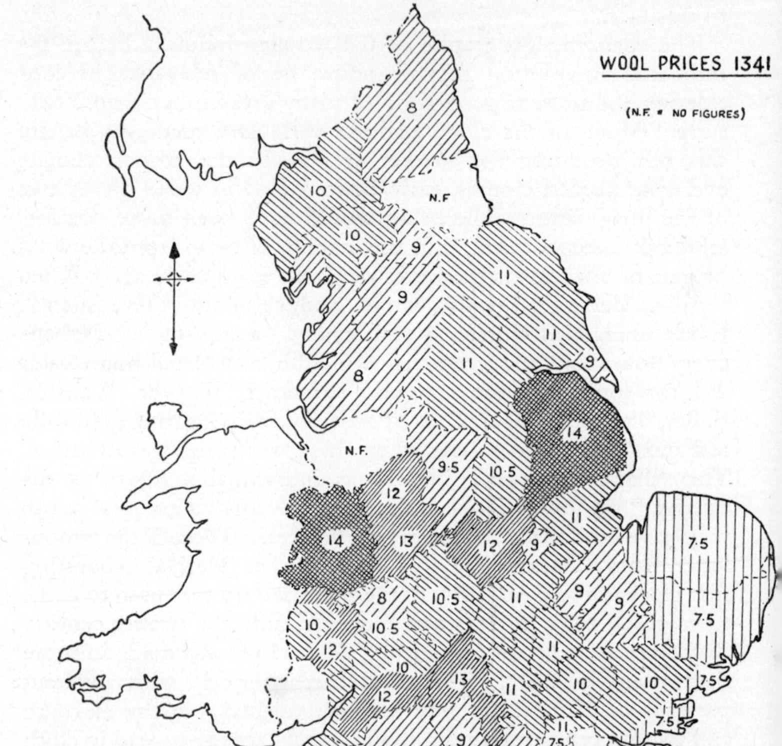

Fig. 41. The figures give the value of the wool in marks per sack. (The value of the mark is 13s. 4d., the weight of a sack is about 28 stones.)

and " Milanese Stamfords of Monza " are listed as customed at Venice. Stamford, it would seem by then produced cloth not only in sufficient quantity to be exported, but also of a quality to encourage imitation abroad. It is difficult to say whether these casual references reflect truly the distribution of the industry in this period. It is probable enough that it would be in towns that production of cloth for market would first occur; here a relatively large number of people are to be clothed, a relatively large number make cloth for a living; thus the chance of a surplus being made, and a surplus of good quality because workers get much practice, is perhaps greater than in a village. Not all towns however seem to have guilds associated with cloth making early or to have been famous for particular cloths. It is impossible to say now what determined the fame of the few but it is perhaps legitimate to draw attention to the fact that most of these towns are near areas which taxation records of the fourteenth and fifteenth century show grew wool of high price and therefore presumably of good quality (Fig. 41). The first essential for fine cloth, fine wool, was available to them.

From the fourteenth to the sixteenth centuries there is evidence that villages in addition to, and presently instead of, towns become leading centres of cloth manufacture. First, towards the end of the thirteenth century the cloth guilds of the towns are crying poor mouth; and in 1304 the weavers of York explicitly state the reason for their distress, " divers men in divers places in the country " are making cloth in competition with them. Secondly, in the manor records of the same date up and down the country a new source of income begins to be mentioned—the fulling mill. The earliest mention of fulling mills yet found in England is in 1185, at which time there are fulling mills on the Templars' estates at Temple Newsham near Leeds and at Barton, new Temple Guiting in Gloucestershire. The distribution of the fulling mills of the thirteenth century cannot yet be exactly established but enough have been incidentally noticed in documents examined for other purposes for it to be clear that already in the villages of the Lake District, the Pennine valleys, the Cotswolds, and Devon and Cornwall, fulling mills were a common source of income (Fig. 42). Professor Carus-

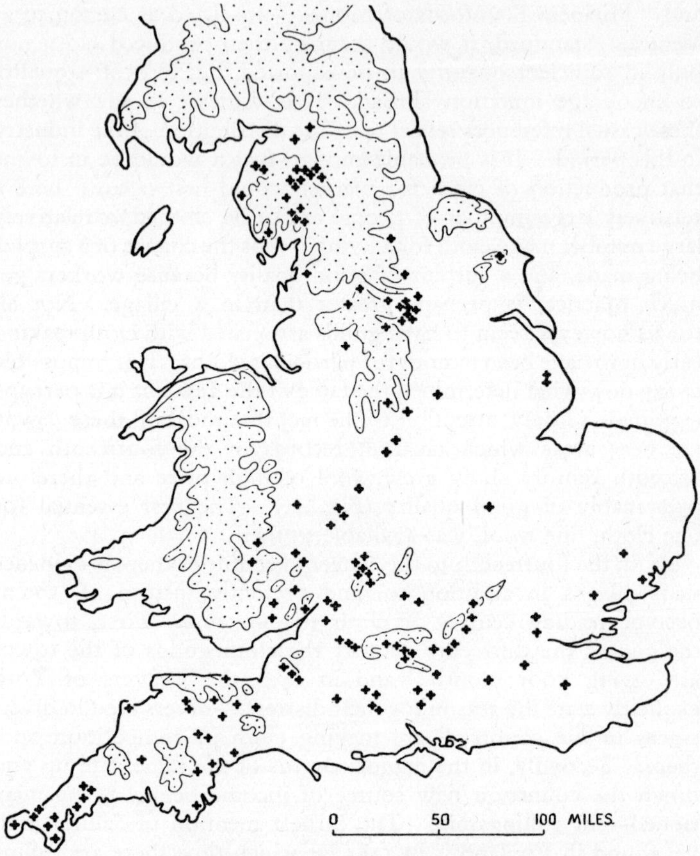

Fig. 42. *Distribution of fulling mills in documents before the time of Edward III. (The map was made from the list given in E. M. Carus-Wilson,* "An Industrial Revolution of the Thirteenth Century." *Econ. Hist. Reviews, XI.*, 1941.)

Wilson goes so far as to suggest that the introduction of the fulling mill brought about a change in industrial pattern sufficient to create "an industrial revolution of the thirteenth century". The process of fulling was a very important one in the manufacture of good cloth. After the cloth was woven it was washed and beaten to remove dirt and surplus grease, to felt the fibres and to ensure that the cloth was " fully shrunk " and thereafter would maintain its size. Clean water was therefore needed and fuller's earth was often added in lieu of soap. At first, fulling was carried out in troughs; light cloths were washed and squeezed by hand, heavier ones tramped under foot—hence the surname " Walker ". It was an obvious move to take the cloths to the river to full them in the running water, and to make some mechanical device to do the heavy work of kneading the fabric. In simple fulling mills or stocks a water wheel turned by the current raised and lowered a row of hammers attached to a beam across the stream. Rivers in towns would rarely be suitable for fulling mills; the river there was often too wide and deep, too muddy and, perhaps most important of all, too polluted. Fulling mills could most easily be erected and worked where the stream was narrow, shallow, with a clean stony bed. The river current turned the wheel, but early mills were small and needed no great power. Soft water fulled cloths best, but clean water was the first essential. The fullers moved out of town to valley villages. Many spinners were already country folk, often women; weaving could be carried on as well in the country as in the town, and perhaps the weavers soon followed the fullers to the country, where living was cheaper and there were no guild restrictions. The dyeing and finishing processes may have remained in town. Dyeing, whether in town or country, had to be done in vats and these used a limited quantity of water which if necessary could be allowed to stand long enough for mud and debris to settle before use. Dye-stuffs were precious; woad (blue), weld, saffron (yellow) and madder (red) were obtained from plants that would grow in Britain, but a considerable quantity was imported from France and Flanders. The most expensive dyes came from Mediterranean ports, vermilion, a crystalline substance found on the shores of the Red Sea, and " grain ", a brilliant scarlet dye obtained from the little insect, *Coccus ilicis*, which lives on the *Quercus coccifera*, the prickly oaks of Mediterranean lands. The alum and much of the potash used as mordants to fix the dyes had to be imported too. The dyers therefore had to be in touch with merchants engaged in foreign trade; there was nothing to take them out of town and much to keep them in town. They needed capital; they were often important citizens, bigger

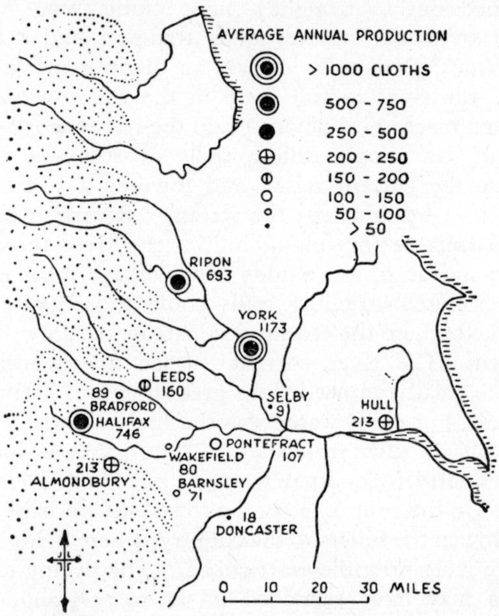

Fig. 43. Cloth Production in Yorkshire in the Fifteenth Century. The figures give the average number of cloths ulnaged in each centre 1473–75. *They suggest that the woollen industry is already well established in the West Riding.*

men than the weavers and fullers. Some cloth was woven from yarn already dyed in the wool, even if it was later dyed again in the say, i.e. after weaving, or after fulling, which was a highly skilled art using expensive dyes and really part of the finishing processes. The other finishing processes, brushing, shearing and pressing were also usually carried out in town. In fact, in the Middle Ages, much of the best quality English broadcloth was exported "white" to be dyed and finished abroad. The cheaper lighter cloths of narrow width, the "kersies and straits", on the other hand, were dyed and finished in the villages where they were woven and fulled. Thirdly, the ulnage accounts suggest a widely distributed country industry with cloth

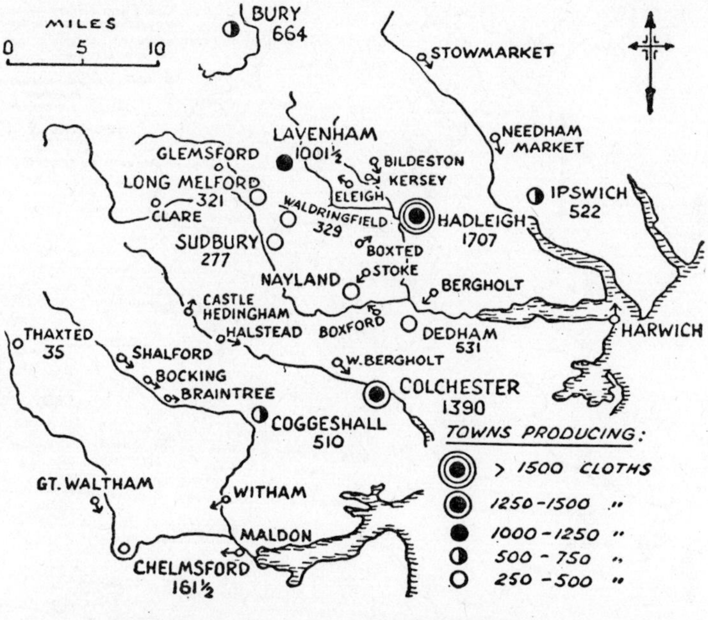

Fig. 44. *Production of Cloth in Essex and Suffolk in the Fifteenth Century. The figures give the number of cloths ulnaged in each centre between Michaelmas* 1468 *and Michaelmas* 1469. *The smaller villages brought their cloth to be sealed to the nearest cloth market as shown by the arrows.*

making already important in the Yorkshire dales (Fig. 43) and still more in Suffolk and Essex (Fig. 44) and Wiltshire and Gloucestershire. Cloth had to be made according to government regulation to safeguard the export market; and cloth before being exposed for sale had to be inspected and sealed to guarantee that it was up to standard. The official who inspected the cloth and later collected the tax upon it was called the

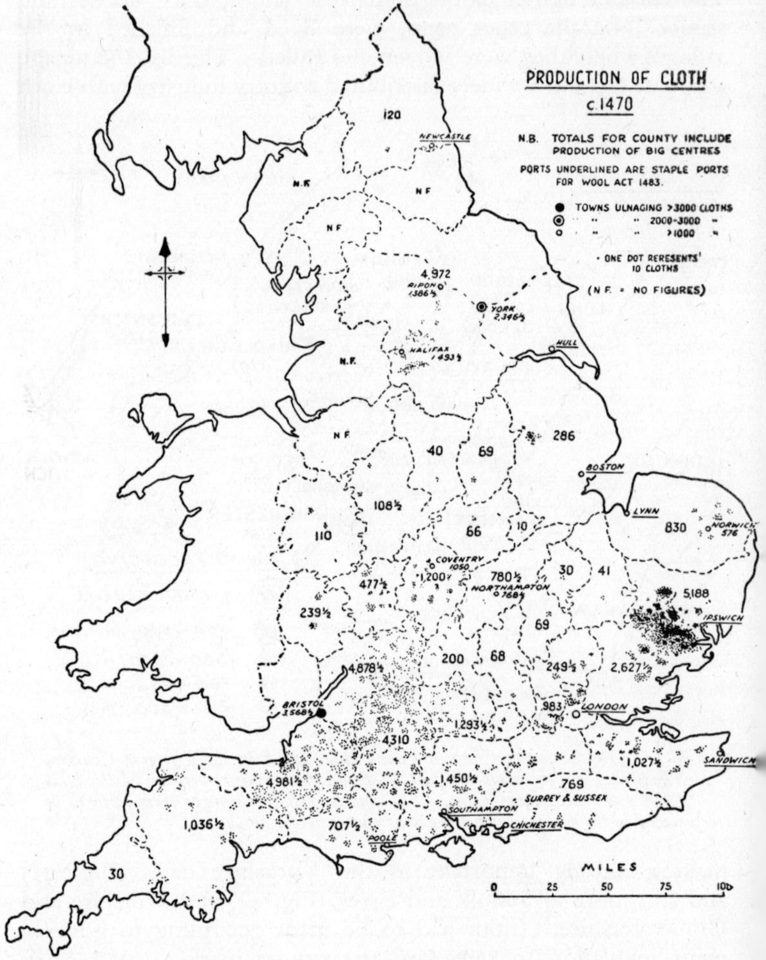

Fig. 45. Cloth Production in England in the Fifteenth Century.

"ulnager", and his accounts, some of which have survived, are known as the ulnage accounts. Figures for most counties survive for several years; one set for the middle, another for the end of the fourteenth century, and yet another set dated about 1470. Fig. 45 shows the distribution of the cloth industry for the later fifteenth century *if* the ulnage accounts can be relied upon. Professor Carus-Wilson has shown that in detail these accounts were often faked. They certainly cannot be used to determine the output of individual clothiers, but probably reflect the general regional distribution of the industry. The greater part of the Gloucestershire cloth is ulnaged in Bristol, pointing to production in and near that city. A list of able-bodied males of military age entitled Men and Armour of Gloucestershire 1608 gives the occupations of most of the men, and in this list in the southern hundreds only—the area in which lies Painswick, Bisley, Stroud, Minchinhampton and Dursley—numerous textile workers are recorded. Here they are very numerous; more than 35% of the men whose occupations are given live by cloth making. The northern hundreds are almost purely agricultural; the greater part of the wealth of the villages of the northern Cotswolds came from wool; Chipping Campden, Moreton in the Marsh, Winchcombe, Stow-on-the-Wold, and Northleach, built their beautiful houses and magnificent churches from the profits of wool sold to the clothiers at the markets of Cirencester, Tetbury and Lechlade. In Suffolk, if the ulnage accounts are used as a guide, the industry rises to importance in the late fourteenth century:—

	1354-58	1394-98	*c.* 1470
Average annual figure of cloths ulnaged in Suffolk and Essex	678	5397	7815

The distribution of prosperity in the cloth-making area can be plotted for various dates from taxation records. If the pattern shown by the records of the value of the ninth in 1341 is considered, the villages later famous for their cloth do not stand out as particularly wealthy (Fig. 46). In the Subsidy Return for 1524 the same centres as those in which large numbers of cloth are recorded as ulnaged stand out from their neighbours. (Fig. 44 and Fig. 47). Local architecture often affords corroborative evidence. The little towns and villages that are recorded as centres for the ulnaging of cloth have very often still to be seen houses and cottages with weaving rooms, and splendid Perpendicular churches with brasses and monuments

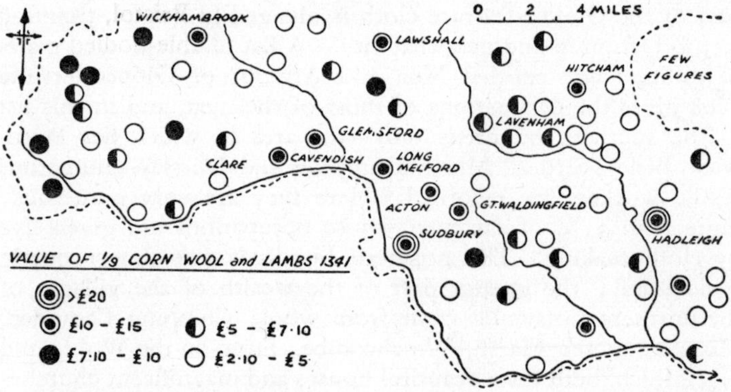

Fig. 46. The Value of the Great Tithes (the ninths) in 1341 in the hundred of Babergh, Suffolk. The villages in which most cloth is recorded as ulnaged in the late fifteenth century (Fig. 44) do not here stand out as particularly wealthy.

of the clothiers who contributed to their building. Wills or Inquisitiones Post Mortem often suggest or prove by their provisions the connection of the testator with the cloth trade, and give his place of residence. Again these in Suffolk confirm the pattern of distribution given by the ulnage accounts.

Many statutes of the fourteenth, fifteenth and sixteenth centuries lay down regulations for the making of cloth. These statutes identify the cloths by name, the names of the counties, towns or villages particularly associated with their manufacture; for example, Kentish broadcloths, Manchester cottons and Welsh friezes, cloth of Kendal, and of Coventry, Pennystone whites and Coxall whites, to cite but a few, are mentioned. These statutes give further evidence of the distribution of the industry (Fig. 48). It is clear from the regulations that different parts of the country made cloths of different kinds: the kinds were of different width, length, weight and colour, varying with the wool used and the technique of manufacture. A great variety of cloths were made; but they fall into three major categories: the broadcloths, of more than a yard in width and usually of fine

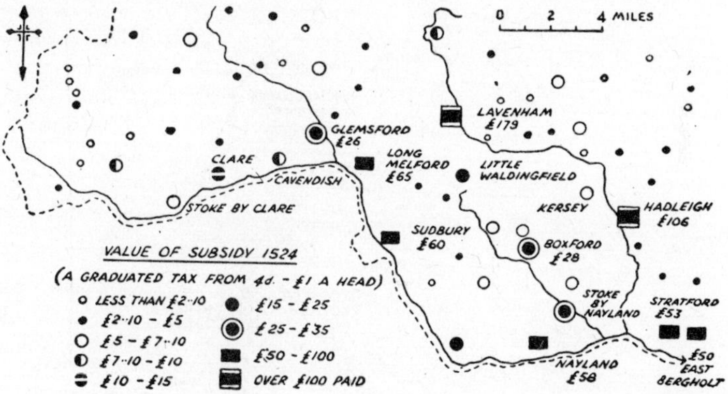

Fig. 47. The value of the subsidy, 1524, *in the hundred of Babergh, Suffolk. The village recorded as ulnaging most cloth* 1468-1469 (*Fig.* 44) *here stand out as particularly wealthy.*

quality; the narrower and lighter cloths, the kersies and the straits, usually cheaper and perhaps coarser than broadcloths; and the worsteds, made from combed wool by less elaborate processes and much cheaper again. The worsteds are always clearly distinguished from the cloths and in the Middle Ages seem to have been little made outside the villages of north-east Norfolk. It is testimony to the one-time importance of East Anglian industry that, although Worstead and Kersey are today unspoilt villages set in a wholly agricultural countryside, their names are used still in the cloth trade, though in the course of centuries the cloths so named have changed again and again in character.

The worsteds did not pay tax when ulnaged and thus are not included in the ulnage returns. They were, however, of sufficient importance to be included in 1347 in the Cloth Custom, a tax paid by denizens and aliens on every piece of cloth or worsted exported. The worsteds paid only a light tax; denizens

Fig. 48. Cloths mentioned in Statutes of the Fifteenth and Sixteenth Centuries.

for example paid 1d. on a single worsted and 2d. on a double worsted but they paid 2s. 4d. on a woollen cloth of full dye and 1s. 2d. on a woollen cloth without dye. The export trade was a considerable one: tax was paid on 29,241 worsteds between Michaelmas 1359 and Michaelmas 1360. Yarmouth and London handled most of the trade and the Hanseatic markets took the bulk of the consignments. Existing evidence suggests that the worsted industry declined in the fifteenth century, perhaps partly in the face of the competition of the cheaper kinds of English woollens, made in increasing quantities at this period in Suffolk and Essex, and in Yorkshire, partly because the manufacture of a worsted type of cloth was taken up at this time too in Armentières and other Flemish centres. Why the worsted technique was adopted in England in Norfolk, and in Norfolk alone, in the Middle Ages is difficult to explain. It is often suggested that worsted was made here because Norfolk sheep had fleeces of long wool. There is little evidence to show that in the twelfth and thirteenth century Norfolk sheep had long wool or that worsteds were made with long wool. The suggestion carries back to the Middle Ages a link that belongs to the present. It may possibly be that *coarse* wool was difficult to card, and the wool of Norfolk sheep, at any rate those reared on the marshland around the Bure valley villages, was probably rather coarse; it certainly was relatively cheap (Fig. 41). The making of stuff from combed wool was certainly a simpler and possibly an earlier process than the making of cloth from carded wool, and may have been the technique first adopted in the Low Countries, where some local sheep too would be marsh fed and therefore probably grew wool coarse and springy in texture. It may have been the Flemings who settled in Norwich soon after the Norman Conquest who introduced the method of manufacture to the district. The sluggish, muddy streams were unsuited to fulling, and thus it is hardly surprising that when the fulling of cloth became a country industry it did not establish itself in the Bure valley, which was left to carry on its own manufacture.

In the seventeenth and eighteenth centuries, technological changes in manufacture, political unrest on the Continent upsetting the foreign market but encouraging immigration of textile workers to Britain, and the increasing strength of trading links with America and the Far East, all combined to bring about a change in the pattern of industry in England. The woollen and worsted industries became more specialised and more strongly localised. War on the Continent and economic policy at home upset the European finishing trade and the demand abroad for the fine West Country broadcloths "in the white" declined; unrest depressed too the market for finished cloths of cheaper quality and they had to compete increasingly with local production in Germany. Settlers in the New World could not afford high-priced cloth, and the trading companies, the Muscovy

Fig. 49. *Cloths mentioned in Thomas Fuller, Church History*, 1655.

Company, the Levant Company, the East India Company, demanded for their markets attractive cloth at reasonable cost. Britain had to adapt her industry; she did it by increasing the diversity of her production and increasing the efficiency of her processes. Cloth to sell had to be well suited to the market demand, and to pay must be made in places well suited to its manufacture. Fuller in his *Church History* (1655) gives a list of cloths made in England which though it purports to refer to the fourteenth century, does refer to his own day (Fig. 49).

It no longer paid to produce fulled cloth in Suffolk and Essex. A progressive increase in the size of the fulling mill involved a change in the physical conditions suitable for the industry. No longer would any little stream with a constant flow of water and a clean shallow bed suffice. New fulling mills tended to be built in the bigger and steeper valleys where a considerable head of water could be engineered; and the small mills on the small streams draining the glacial drifts of the East Anglian Heights could no longer produce economically, especially with their foreign markets upset by war. The clothing villages of Suffolk and Essex by the later decades of the sixteenth century were in a bad way. But Continental wars did good as well as harm; many cloth workers fleeing from the wrath of Philip II took refuge in England. Many settled very naturally in cloth-making districts, and in those where native fabrics were no longer a paying proposition they introduced most successfully "new draperies". These "new draperies" were made from combed wool and did not need fulling. They had many outlandish names: bays, says, serges, fustians and velours soon became familiar English words and have remained so, though with many a varied meaning. But the bewpers, and the buffins, the cruells and the dornicks, the mockadoes and minikins, the pomettes and perpicuanas, the rashes and the stamells, the tukes and the tamettes and all the many more that tripped so lightly from the pen of seventeenth-century officials leave us flummoxed. It is impossible to imagine exactly their appearance or their uses, but it is also impossible to fail to be impressed by their astonishing variety. Sixteenth-century immigrants settled in Colchester and Sudbury; and in the villages of Essex and Suffolk bays and says soon ousted kersies and straits. The bays and says seem to have been at first a relatively coarse cheap cloth, but trade in them became sufficient to revive and maintain prosperity in the area, as the Queen Anne and Georgian houses that still today grace the streets of Clare, Long Melford, Sudbury, Lavenham, Hadleigh, Dedham, Nayland, East Bergholt and the rest show conclusively.

Norwich too attracted Dutch and Walloon settlers, and soon a great variety of " new draperies " were made in and around the town. There were the camlets, workaday stout cloths much used by Spanish and Italian monks for cloaks, and for raincoats by the British ; there were the shalloons " worn by the lower class of females " ; there were the fancy striped cloths used for upholstery and the finer varieties for dresses ; and there were the mixed fabrics of silk and wool, the bombazines and fine damasks still so much beloved by the heroines of Jane Austen and Mrs. Gaskell. The old medieval name " worsteds " became applied to these new cloths, presumably because they were made from combed wool, but they can have had few other characteristics in common. Norwich soon eclipsed Suffolk, and needed more wool than Norfolk could supply ; fine wool of long staple, the best kind for combing, was imported from Leicestershire and Lincolnshire and was put out to spin to villagers not only in Norfolk but in Suffolk too. In fact Suffolk's prosperity came to depend to a considerable extent on supplying yarn to Norfolk weavers. Defoe in 1724 describes :

> "a Face of Diligence spread over the whole country: the vast manufactures carried on by the Norwich weavers employ all the country round in spinning yarn for them ; and also many thousands of packs of yarn which they receive from other countries even as far as Yorkshire and Westmorland. . . . This side of Norfolk is very populous and thronged with great and spacious market towns, more and larger than any other part of England so far from London, except Devonshire, and the west riding of Yorkshire ; for example between the frontiers of Suffolk and the City of Norwich . . . are the following market towns, viz.

Thetford	Hingham	Harleston
Dis	West Deerham	East Deerham
Harling	Attleboro'	Watton
Bucknam	Windham	Loddon etc.

> Most of these towns are very populous and large, but that which is most remarkable is that the whole country round

them is so interspersed with villages, and those villages so large and so full of people, that they are equal to market towns in other counties."

Norwich stuffs were held in high repute at home and in Europe, and Norwich obtained the monopoly of the East India Company's market in the Far East.

In the West Country, the narrow valleys with a good fall of water which drain the Marlborough Downs, Salisbury Plain, the Mendips and the Southern Cotswolds provided a favourable physical setting for the economic manufacture of cloth well fulled in up-to-date mills, and therefore the clothing areas of west Oxfordshire, north Wiltshire, north Somerset and south Gloucestershire retained pre-eminence in the manufacture of broadcloths of first quality. On the south-eastern edge of this area the old medieval industry of Berkshire and Hampshire decayed, but the south-western fringe took up the manufactures of new cloths. In the Parrett valley and in south Devon, the Somerset and Devon serges gained a great reputation. They were mixed fabrics with a worsted warp and a woollen weft. Celia Fiennes visiting Exeter in 1698 says:

"as Norwich is for copes, callamancos and damask so this is for serges—there is an incredible quantity of them made and sold in the town; . . . the whole town and country is employed for 20 miles round in spinning, weaving, dressing, and scouring, fulling and drying of the serges, it turns the most money in a week of anything in England".

Defoe in 1724 considers that the weekly serge market in Exeter is "very well worth a Stranger's seeing, and next to the Brigg market at Leeds in Yorkshire, is the greatest in England".

Yorkshire, more remote, less populous, and less well endowed with fertile soil than south-western and eastern England, was steadily developing her manufactures. A writer in 1561 points out the advantages of the dales:

"in Halifax, Leeds, and Wakefield for that not only the commodity of the water mills is near at hand but also poor folk as spinners, carders and other necessary workers for the webbing may there . . . have rye, fire and other relief good cheap "

The Yorkshire streams afforded ample power for fulling mills, and the output of woollens of a great variety, and some by now of very high quality, increased steadily. Defoe states that " cloth (i.e. broadcloth), kersey or shalloon . . . are the three articles of this country's labour ". He gives a vivid picture of the thickly scattered population of the upper dales

" so still the nearer we came to Halifax, we found the houses the thicker, and the villages the greater in every bottom; and not only so, but the sides of the hills which were very steep in every way, were spread with houses. . . . We found the country one continuous village though in every way mountainous, hardly an house standing out of speaking distance with another; and as the day cleared up we could see at every house a tenter and on almost every tenter a piece of cloth kersey or shalloon ".

He draws a contrast with the lower land :

" From hence (Burstall) to Leeds and every way round the inhabitants appear exceeding busy and diligent : the houses are not scattered and dispersed but crowded up in large villages and thronged with people ".

Of the cloth market at Leeds he writes :

" a prodigy of its kind and perhaps not to be equalled in the world. . . . Whatever stranger happens to be in Leeds on a Tuesday or Saturday should not omit the seeing of this incomparable market. . . . Thus you see 10 or 20,000£ worth of cloth and sometimes much more bought and sold within the hour ".

He tells of the extensive trade with neighbouring counties and makes the point, " thus one trading manufacturing part of a county, in a barren soil, gives and receives support from all countries round it". Yorkshire villages were well situated for the manufacture of cloth and for long remained fully occupied by it and they paid little attention at first to the newfangled methods of manufacturing of the "new draperies". But Yorkshire became drawn into this industry too: Norfolk's demand for wool reached Yorkshire and Yorkshire sheep farmers found a market for their clip in Norwich; then Yorkshire took to combing and spinning the wool before export and sent worsted yarn to Norwich. Yorkshire wool was not of the longest staple and thus not best suited to the making of worsted yarn. She began therefore to import fine long wool from Leicestershire to spin into these yarns for Norwich industries. Finally she took to using some of the worsted yarn to make worsted cloth in Yorkshire, at first trying her hand at the coarser and cheaper qualities. Thus were set the conditions which enabled her industries to learn from the cotton industries the benefits to be gained from mechanisation, rather than to shrivel and decay under the strain of competition.

The Changing Geography of the Iron Industry

Early iron industry leaves few permanent traces. Iron mines and "hammer" ponds may be but small dimples on the ground. Works were small scattered hearths, and workers, if something more than itinerant smiths, seem to have been men of few possessions. Iron-masters perhaps lived elsewhere. Certainly early iron-working districts are not distinguished by imposing guildhalls, merchant houses or magnificent churches; nor are they marked by villages larger than an agricultural area would seem to warrant. Early iron industries often existed in "the wilderness" and left little trace when conditions changed and the industry passed away, and thus they are unimportant for the legacy they leave.

Later iron industry on the contrary scores an area indelibly. Once ore and coal are extensively mined, once blast furnaces, steel works and rolling mills are established, mining villages and iron and steel towns grow up which, with the investment of much capital in immovable and expensive equipment and the development of a peculiar economic and social tradition, create a strong geographical inertia. Regions of heavy industry show a pattern difficult to change and impossible to obliterate. Therefore the conditions that originally favoured the growth of the industry demand close investigation since, although they may no longer operate directly, they may play a very important part indirectly in present designs.

But if patterns now change slowly, they do change, and once changed rapidly. Thus the geography of the iron industry of successive periods affords an excellent opportunity to examine the changes in geographical patterns brought about by changing evaluation of geographical conditions in response to changing supply of raw materials, changing techniques of manufacture, changing products and changing markets. It is from this point of view that the main facts of the geography of the iron industry are briefly considered here.

Sources for the study are scanty for the earlier periods but abundant for the eighteenth and nineteenth centuries. References to iron works in official and manorial records of medieval and Tudor date are sufficiently abundant to establish in general the iron-working districts, but tell little of the position of the furnaces and forges within these areas. Finds of cinder heaps, hammer ponds, and even of the ruins of mills themselves, may help to establish the distribution of early industry within any one district. But they are difficult to identify and still more difficult to date. Cinders have become much scattered; by the plough, and because in some areas, the Weald clay vale for instance, cinder was a much prized road metal. Cinders are hard to date; the extraction rate did not necessarily change much from period to period, it often varied from district to district, and early cinder heaps were often re-worked at a later date. Hammer ponds, so called, may be hollows on the site of the pits from which ore was dug, hollows dug or valleys dammed to contain water to wash ore, to cool hammers in the forge, or to turn wheels to blow bellows or work hammers. They may also be old marl pits, perhaps quite recently made by eighteenth-century "improving" farmers, or fish ponds, or mill ponds belonging to other mills than iron mills, and thus not associated with iron industry at all. Field names may be useful but again they are difficult to interpret in detail and to date. For the later centuries, descriptive works and statistical sources become increasingly numerous and reliable; many of the latter allow distribution maps of fair accuracy to be constructed.

Four major periods preceding the present may be distinguished. In the first, from earliest times to the early eighteenth century, ore was smelted, until about 1500 on a bloomery hearth, later in a blast furnace, and refined in forges, using charcoal as fuel at all stages in the manufacture. Industrial units were small, products were small, and the industry widely scattered; the monks of the Pennine and Cleveland monasteries worked iron, and the Forest of Dean and the Weald were particularly famous iron-making areas. This was the Age of Charcoal. In the second period, roughly coincident with the middle decades of the eighteenth century, iron was smelted with coke fuel in the blast furnaces but was refined in forges still using charcoal fuel. Water power was essential to work the bellows of the furnaces and the hammers of the forges, now of a considerable size. The well-wooded river valleys of the Midland coalfields became the leading iron-producing district. This was an age of transition. In the third period, in the late eighteenth and first half of the nineteenth century, with the introduction of the steam engine and of the puddling process by which iron could be forged with coke, the industry was freed from dependence on water for power and wood for fuel; it became closely tied to the coalfields of the Midlands, South Wales and Central Scotland. This was the age of coal. In the fourth period, the late nineteenth century, the demands of the markets for steel and the limitations of the first cheap steel-making processes gave a great incentive to iron industries with easy access to non-phosphoric ores, to Cumberland and Furness which possessed the only suitable ores for steel making in Britain, and to the North-East with a well-established import trade in iron ore. In the age of steel, a coastal situation was an added asset since Britain made so great a part of the world production of iron and steel and supplied so many overseas markets; Tees-side dominated the industry.

The Age of Charcoal

In the early period, easily mined ore, abundant wood for charcoal, and easy transport to market were of first importance. The relative importance of these factors varied: the period may be roughly divided into two, an earlier one when wood was plentiful wherever there was ore, so suitable ore to smelt was of first importance; and a later period when wood was scarce and the existence of a cheap fuel supply was of primary importance; ore was taken if necessary to fuel. In the earlier phase a further division can be made; when iron was smelted on a bloomery hearth the temperature was not great enough to melt the metal and a malleable mass of iron collected on the saucer-shaped hearth. This mass was often very impure and was hammered to get rid of the clinker, and then wrought into small bars or small iron products, arrow heads, horseshoes, nails and bolts, ploughshares, cart wheel rims and so forth on the spot. These small products were easy to transport, so that distance to market in terms of mileage usually decided where, given iron, an industry would develop. About 1500 in the Weald, about the middle of the century in the Midlands and at the beginning of the seventeenth century in the Forest of Dean, the blast furnace was introduced. The iron and wood were built up much higher within a cylindrical containing wall and a much stronger draught blown through than in the old bloomery process. The temperature generated melted the iron which was led out in a liquid stream from the base of the furnace to cool in a sand bed with a central trough (the sow) and little lateral bays (the pigs). This iron was hard but brittle with a considerable carbon content; it was either reheated and run into moulds of the shape required, cannon balls, cannon barrels, grave slabs, firebacks and the like, or it was heated and hammered on a finery hearth until the carbon content was reduced and the iron became malleable. It was then made into bars or usable objects. The size of the blast furnace needed powerful bellows which needed water power to blow them; the heavy work in the forges to hammer pig-iron into wrought iron needed hammers worked by water power; furnaces and forges were built in the upper valleys of considerable rivers where dams could be easily made to make water shutes to turn wheels. More fuel was needed in the forges where pig-iron was made into bars of wrought-iron, than in the furnaces where the raw ore was made into pig-iron; so whenever objects could be cast directly into a mould from liquid pig-iron this was done. Cast-iron products, particularly cannon, were big and heavy and cumbersome to transport. They could no longer be rafted down any little stream, or carried on packhorse across any surface. Only the bigger and deeper rivers would float boats to carry them, and as the need for water power took furnaces and forges upstream the problem of bringing the products to navigable water increased. Now not only iron and wood for charcoal to smelt it sufficed for a flourishing industry, transport became a key factor; both means to carry ore and fuel to the nearest suitable stream which would work bellows and hammers for furnace and forge, and means to carry the finished products to market were needed. Ore was heavy in relation to its value and thus would not bear cost of transport very far; the products were heavy, cast-iron products were often too heavy for pack horses, they needed carts or boats. Nearness of ore to iron-works or of iron-works to market could no longer be appreciated in terms of mileage only, ease of transport mattered more than distance.

The Forest of Dean seems to have had in the early centuries a more important iron industry than the Weald. It had iron ore irregularly deposited in the Carboniferous Limestone, and the iron-works mentioned in a verdict on Forest works about 1250 are most markedly arranged in a wide circle round the Forest along the outcrop of Millstone Grit. Wood was abundant and the numerous little streams running from the Forest highland to the Severn and the Wye provided all the water needed. The smelting was done as near as possible to the ore for it was easier to carry the finished products. The Forest of Dean iron-works lay nearer than the Wealden ones to the main market. Iron was most used by the army; the army needed arrow heads, nails and bolts for chariots and fortifications, and horseshoes, only warhorses were shod. The troubled frontier in late Saxon and early Norman times was in the West; Mercia and Wessex were pushing west against Wales, and sometimes clashing with one another. Mercia, Wessex and perhaps Wales too, drew on Forest of Dean iron-works for equipment. The armies of the Norman and early Angevin kings were engaged in border wars against the Welsh and campaigns in Ireland.

In the late thirteenth century and in the fourteenth century there are more frequent references to iron-works in the Weald than in the Forest of Dean. The King appears to be buying for his army here: in 1254 the Sheriff of Sussex is asked for 30,000 horseshoes and 60,000 nails, in 1275 the King's chief smith buys 406 iron rods for £16 17s. 1d. "in the Weald"; in 1320 the Sheriff of Surrey and Sussex furnished horseshoes and nails for Edward III's Scottish expedition. There are few records for the fifteenth century but numerous surveys in the sixteenth century show the industry well established and in 1572, according to a complaint of "wood despoiled and cannon exported", there were "not so few as 100 furnaces and iron mills in Sussex, Surrey and Kent." The first list of them is dated 1574 but it is a list of the owners of the mills, not of the places where they are, and the position of some of them is doubtful. Throughout the sixteenth and seventeenth centuries, however, the Weald was undoubtedly the most important centre of iron working, the main ordnance factories of the day were here, and much bar-iron was exported to the towns to be made up at the smithies into products in demand.

The Weald had, like the Forest of Dean, suitable ore near the surface. A clay ironstone occurs in nodular or tabular blocks at the base of the Wadhurst Clay, the middle bed of the Hastings Sands series; septaria of deep red ironstone occurs in a band two or three feet thick near the upper surface of the Weald Clay itself; and irregular bands of ironstone, known as carstone, occur in the Folkestone beds of the Lower Greensand, especially in Surrey and West Sussex. Early there was an abundant supply of wood for charcoal, and the small streams of the Forest Ridges and of the inner edges of the Downs provided enough water for washing, cooling, for power and even for transport. The Weald had an advantage of position compared with the Forest of Dean once the armies were active on the Scottish border and in France, rather than on the Welsh border and in Ireland, and were therefore supplied from London. The Weald was better placed, too, to supply iron to the London and Winchester blacksmiths to make the bolts and bars and hinges the builders needed; it was in eastern England around the Weald, rather than in western England around the Forest of Dean that demand from civilians was greatest.

At the end of the seventeenth century the Forest of Dean was once more gaining in importance at the expense of the Weald. The Forest had two advantages, better communications and a cheaper fuel supply. Communications in the Weald were good enough to serve the industry in early times: but after the introduction of the blast furnace suitable sites were more limited: the furnace might be some distance from the mine, and often the furnace and the forge had to be some distance apart as one site had not enough water to work both the bellows and the hammers. Wood was less plentiful and charcoal also had to be brought from an increasing distance. The finished products, especially cast-iron products, were heavy and unwieldy and needed good roads and big rivers to transport them. The demand for good communications thus increased, but the occurrence of iron ore and the need for water power concentrated the industry deep within the Weald around the central ridges. The wide band of sticky Wealden clay with muddy roads thus lay between the furnaces and forges, on the one hand, and the navigable reaches of the streams in the lower valleys or the dry roads of the chalk on the other. Heavy loads needed heavy carts with iron-bound wheels and these cut up the roads. In 1743 John Fuller an iron-master writes:

> "I have gotten twenty 9 pounders of 9 feet to Lewes . . . these 20 have torn the roads so that nothing can follow them and the Country curse us heartily".

An army in a hurry for guns, or even an architect wanting his railings, is not going to be well satisfied with such a state of transport. The waterways of the Weald were "improved" and canalised but the rivers were small and water low, especially in a dry season. Although the iron-works of the Weald were so near London in miles, it often took less time and less money to bring goods from the iron-works of the Forest of Dean. The roads over the Millstone Grit and Carboniferous Limestone were dry; the Wye and the Severn and many of their tributaries were navigable; and transport by sea to London cheap.

It is frequently stated that the decline of the Wealden iron industry was the result of the *exhaustion* of the fuel supply as a result of the *disappearance* of the wood. The fuel supply was not exhausted, the wood did not disappear; but increasing demands for wood led to increasing scarcity, first of log wood and then of coppice wood for charcoal, and thus to increasing cost of fuel. What did disappear was fuel at an *economic* price. The iron-works were blamed for the scarcity of wood; a timber surveyor in 1574 writes:—

> "It may please your honour to consider the several notes ensuing which do concern the great spoil and consumption of Okes timber and other woods which occur in the counties of Sussex and Surrey and Kent by means of the iron mills and furnaces: unless speedy remedy be provided there shall not be timber sufficient to be had within these few years for Her Majesty to build any shippes or otherwise".

The surveyor may have exaggerated the seriousness of the position but it was true that wood was becoming scarcer and it was true that the iron mills used a lot. A proposal to Parliament in 1559 that iron mills should be banished from the Realm stated, "where wood was formerly sold at the stock for 1d. a load by reason of the iron mills it is now 2s. a load". No one was harder hit by increased cost of wood than the iron-masters: it is reckoned in Henry VIII's reign that the cost of fuel represented 65% of the cost of producing pig-iron from ore at the furnace, and 51% of the cost of converting pig-iron to bar-iron at the forge; and in 1746 Fuller states that the cost of fuel forms 83½% of the cost of moulding and casting guns. The advantage of the Forest of Dean, with its greater reserves of wood within the Forest and the possibility of importing wood from the forests of Brecknock, Hereford and Radnor, was obvious. The store here was less depleted since population was less dense and industries of all kinds less numerous and well developed than in south-east England.

Even the Forest of Dean had not an unlimited store of wood and in the later decades of the sixteenth century and the early decades of the seventeenth century the iron industry "takes to the wilderness". There is an increasing tendency to set up furnaces, and even more especially forges, wherever abundant wood and therefore cheap fuel exists near navigable water. It was abundant wood that played a big part in building the iron industry of the Midlands. The woods, the Wyre Forest, Corfe Wood, Norwood, that clothed the valleys of the Severn and the Stour were remote and relatively undepleted and in the seventeenth century much pig-iron was exported up the Severn to be converted into bar-iron. Andrew Yarranton in 1677 tells us:

> "the greater part of this Sow iron is sent up Severne to the forges, into Worcestershire, Shropshire, Staffordshire, Warwickshire, and Cheshire, and there it's made into Bar Iron: And because of its kind and gentle nature to work, it is now at Stourbridge, Dudley, Wolverhampton, Sedgley, Wasall, and Birmingham and thereabouts wrought and manufactured into small Commodities and diffused all England over . . . and when manufactured sent into most parts of the World".

There was of course local iron in the Midlands, but as Yarranton tells us it was "of another nature quite different from that of the Forest of Deane. This Iron is a short soft Iron, commonly called Cold-shore Iron, of which all the nails are made and infinite other Commodities". Yarranton estimates that there are 60,000 persons employed in iron working in the Forest of Dean, and in the Midlands "many more persons if not double to what are employed in the Forest of Deane". His figures may be totally unreliable but it is clear that in the Midlands at the end of the seventeenth century iron industry was of great importance.

Hunger for fuel forced iron-masters to more and more remote lands. They migrated to Derbyshire and South Yorkshire where local ore and abundant wood favoured the growth of industry along the Don and its tributaries around Worksop, Chesterfield, Sheffield and Barnsley. They settled in Durham and Northumberland; in 1682 Ambrose Crowley, an ironmonger of Greenwich, set up works at Sunderland and employed many foreign workers, in particular men from Liége, and in 1690 moved his works to a still less populous region, to Swalwell and Winlaton near Newcastle, where wood was plentiful and coal near at hand. There was no smelting at all in this region before 1745, Swedish and American pig-iron was imported and worked into bars in in the forges, turned into sheets, cut into rods in the slitting mills, wrought into anchors and chains and pumps at Swalwell; into nails, files, chisels, saws, hammers and agricultural implements at Winlaton. Coal could not yet be used in the forging of bar-iron, but it was already used in the smithies where bar-iron was wrought into finished goods, so this area with abundant wood and coal, with an easy import of first-class pig-iron had already like the Midlands laid in the late seventeenth century the foundations of later industry. The north-west of England also received its iron workers driven north in search of cheap fuel in the early eighteenth century. A pioneer furnace was established at Clifton between Cockermouth and Whitehaven to smelt the local kidney ore. One James Spedding established a forge at Whitehaven and by the mid-eighteenth century others were established at Workington. In 1711 the Blackbarrow Company was established in Furness and by 1748 eight blast furnaces were at work there.

The demand for wood for cheap charcoal grew faster and faster, and not even regions in the west and north of England where there was local iron could provide sufficient cheap fuel. The areas around Whitehaven and Furness were not heavily forested, and once the nearer woods were exhausted the relief made transport of wood from inland too costly. Further and further into the wilderness went the industry. In 1727 partners of the Blackbarrow Company joined with James Spedding and set up industry in Invergarry in Glen More. Here there was no local ore, ore was carried by sea from Furness; no local work-people, masons were brought from Edinburgh, charcoal burners from Ireland, furnacemen from Yorkshire and South Wales. The only advantage was great woods that could be reached by water. These works had a short life; as soon as wood near at hand was cut down and fuel less cheap, the remote situation was an overriding disadvantage. The works at Invergarry closed in 1736. But even after the turn of the century the pull of cheap charcoal took the iron industry to the Highlands. The Newlands Company, failing to gain a footing in Furness, established works at Bonawe on Loch Etive in 1752; in 1775 the Duddon Company lighted a furnace at Inverary on Loch Fyne; and it was originally the attraction of cheap wood that led J. Roebuck, the Sheffield iron-master, to found in 1760 the works to become famous in a later period at Carron near Falkirk.

In the chase for cheap wood Scotland and the north of England came to know industry but never played an important part; the existence of the remote iron-works are interesting merely as an extreme example of the thesis that, just before the establishment of a successful coke-smelting industry, iron went to wood. The list of iron works with their output, published in a pamphlet, *The Interest of Great Britain*, 1721, for all its imperfections probably gives a fair indication of the distribution of the iron industry at the end of the Age of Charcoal (Table I).

The Midlands had already obtained the lead, because of cheap charcoal, not, as so often stated, because of cheap coal; coal was not yet used to make bar-iron. The Forest of Dean was falling behind; it was short of wood and beginning to be short of iron ore. Of the new areas Sheffield, Chester and North Wales and South Wales were of most importance, there ore and wood were found together but poor communications limited the growth of industry. It was abundantly clear that the day of the dominance of the Weald had passed. The acute shortage of fuel even caused a decline in the total output of iron at home in the early eighteenth century. The total output of bar-iron in this country in 1720 was about 20,000 tons, but between

TABLE I

The list of Furnaces and Forges in *c.* 1720 cited by the author of *The Interest of Great Britain.* (Published by D. Mushet. Papers on Iron and Steel. 1840.)

District	Furnaces	Forges	Output of Forges in tons of bar-iron
South East: Kent, Sussex, Hants and Berks	15	12	740
South West: Glos, Hereford and Monmouth	11	20	3950
Midlands: Shropshire, Worcs, Warwick and Staffs.	12	45	7910
Chester and N. Wales: Chester, Flint, Denbigh, Montgomery, Cardigan	5	12	1800
South Wales: Brecon, Glam, Carm, Pembroke	5	10	1370
Sheffield District: Yorks, Derby, Notts.	11	18	2680
North West: Lancs, Cumberland.	—	6	630
North East: Northumb, Durham	—	5	505

1730 and 1750 it was probably less. In 1740 it is calculated that about twice as much iron was imported as was made at home.

The Age of Transition

The shortage of wood not only sent men to hunt for it in remote places, but also set them to hunt for a substitute for charcoal as fuel. The idea of using coal was not new in the eighteenth century. Coal was used industrially from very early times. It was used to evaporate salt from sea water on the Northumberland and Durham coast, and for lime burning wherever conditions brought lime, coal and builders together, as in the London area where lime from local chalk beds and coal imported by sea from Newcastle were used to make plaster for the walls and roofs of an ever-increasing number of houses. It was also early used in the smithies to heat the bar-iron sufficiently to allow it to be hammered into the product required. Three things limited the use of coal: the cost of transport, it could only be used near the coalfield or near navigable water; the dirt and smoke, it could only be used out of doors unless elaborate chimneys were constructed; and the impurity of the product. Coal had much sulphur and phosphorus combined with its carbon and these impurities would in some industries react unsuitably with the material with which it was brought into contact. This last difficulty was the one that long baffled the iron-makers. In 1595 Sir Robert Cecil held a licence to smelt iron with coal; in 1611 a patent was granted to William Sturtevant to smelt as well as to manufacture iron with coal, but his verbose *Treatise Metallica* does not tell *how* he did it, perhaps because he could not do it successfully. In 1621 Lord Dudley took out a patent for his son Dud Dudley who had thrown up his studies at Oxford to make money by his "great idea". Dud Dudley claimed that his method of smelting iron with coal was wholly successful technically and that his economic failure was entirely the result of the jealousy of his betters. Whatever the truth of Dudley's claim, Robert Plot in his *Natural History of Staffordshire*, 1688 states that "Nature never intended the union of coal and iron to be fertile for mankind". Early in the eighteenth century iron was at last successfully smelted by Abraham Darby of Coalbrookdale with coal which he first burnt to remove some impurities. Abraham Darby's success was in part the result of luck. The coal he tried was good for his purpose. The "clod" coal of Coalbrookdale occurred near the surface, was particularly pure, and made good-burning, hard coke. Abraham Darby may have smelted iron successfully with coke about 1709 but it was not until the 1730's and 40's that the process became at all well known or widely taken up. All difficulties were not at once overcome; in 1762 the Society for Encouragement of Arts and Manufactures was still offering prizes for pig-iron made with coal "equally good with that made with charcoal". Coke-smelted iron was superior to charcoal-smelted iron for making cast-iron goods: the higher temperatures produced meant that the iron was more liquid and more even in consistency and could be run in the foundry into finer moulds. But coke-smelted iron was less good than charcoal-smelted iron for making bar-iron: the conversion of the coke-smelted pig-iron to bar-iron consumed even more charcoal than was needed to convert the charcoal-smelted pig, and the finished product had less ductility and tensile strength. For a time therefore charcoal was used to smelt iron intended for conversion to bar-iron, and until the end of the century iron forges were tied to the woods.

The use of coal as coke in the blast furnace gave a great stimulus to the use of cast-iron: it could be made without expensive fuel and it eliminated the cost of the highly skilled work of the men at the forges. There was thus a fight between the two collateral branches of the iron industry, and in the latter half of the eighteenth century the foundry was winning hands down over the forges. In 1751 a patent was granted to make gun carriages from cast-iron, in the 1760's and 70's there were numerous patents allowing coffin nails, then all nails, iron hinges, and buttons to be cast. There was even a patent to allow hoes for West Indian plantations to be made from cast-iron, although they had to be strengthened at the tip with malleable iron. The first iron bridge, erected at Ironbridge across the Severn Gorge in 1779, was of cast-iron; and cast-iron began to replace wood, copper and brass in many implements and machines used in the textile and metal industries.

The late eighteenth century was an age of transition in the geography of the industry too: the forges were still widely scattered and primarily located by the presence of wood for charcoal; but the furnaces and foundries were freed from the forests. They were still tied to river valleys, but to river valleys in or near the coalfields. In 1778 the amount of pig-iron made with charcoal was still considerable in the Forest of Dean, but the amount of pig-iron made with coke in Britain as a whole was more than seven times that made with charcoal (Table II 1798).

TABLE II

Tons of Pig-iron Manufactured in 1788

Districts	Smelted by Charcoal		Smelted by Coke	
South East: Weald	Sussex	300	—	—
South West: Forest of Dean	Glos, Mon,	2600 2100	—	—
South Wales	Glam, Merion, Carmarthen	1800 400	Glam, Brecknock	6600 1600
Midlands	Shropshire	1800	Shropshire Staffs, Cheshire	23100 4500 600
Sheffield District	Derby Yorks,	300 600	Derby Yorks,	4200 4500
North West	Lancs, Cumb, Westmorland	2100 300 400	Cumb,	700

Figures given in D. Mushet, *Papers on Iron and Steel*, 1840.

The Midlands, in particular Shropshire, was the leading region of production. In 1796 there were 22 furnaces along the seven-mile stretch of the Severn north and south of Ironbridge. South Staffordshire, along the Stour valley, south Yorkshire and north Derbyshire along the Don valley had also become important centres of iron-smelting. All these areas had, as Yarranton remarked, " an infinite of Pit Coals and the Pit Coals being near the Iron, and the Iron Stone growing with Coals ". But the true coal age, the complete dominance of coal in the industrial geography of the nation was still to come.

The Age of Coal

In the period 1790 to 1850 the iron industry in all its branches was freed from bondage to woodland for fuel, and, along with other industries, from bondage to running water for power. It was tied more closely than ever to certain coalfields.

Three lists printed by H. Scrivenor in his *History of the Iron Trade* and extracted by Scrivenor from figures given to the House of Commons show sufficiently for the present purpose the distribution of the iron industry in the first half of the nineteenth century (Table III).

TABLE III

Tons of Pig-iron Manufactured in 1806, 1830 and 1852

Districts	1806	1830	1852
Monmouth	2444		
Gloucester	1629	—	—
South Wales	75609	277643	666000
Shropshire	54966	73418	120000
S. Staffs,	49460	212604	725000
N. Staffs,			90000
Derby	10329	17999	150000
Yorks,	26671	28926	
N. Wales	—	—	30000
Lancs,	2500	—	—
Cumberland	1491	—	—
Northumberland	—	5327	35000
Durham	—		110000
Scotland	23240	37500	775000
Total	248331	653627	2701000

Figures given in H. Scrivenor, *A Comprehensive History of the Iron Trade*, 1854.

Incidentally Scrivenor gives details of the production of each furnace for 1823 and 1830, and it may be of interest to those readers resident in iron-making districts to plot these local figures. In general the points to notice in comparing these three lists with that of 1788 and with each other are firstly the great and steady increase in the amount of iron made, and secondly the changes in the areas most important in production.

The great increase in the "make" of iron in this period largely reflects the demand created first by war, largely for cast-iron, and then by the railways, largely for wrought-iron. The wars of the French Revolution and the Napoleonic Wars increased the demand for weapons and shot, and furthermore by upsetting the European production of iron they left Britain without a serious competitor in world markets. By 1831 Lardner over-optimistically writes, "happily the business of cannon casting on a large scale appears to be at an end". New uses of iron however were increasing every day, cast-iron was needed for gas- and water-mains, for buildings, even iron paving was tried in Leicester Square, but above all wrought-iron was needed for railways. The demand fluctuated with spells of railway development but increased not only with the building of new lines but also because later lines were laid with greater weight of metal. From 1831 to 1841 it is estimated that there was an increase from 53½ tons to 156 tons of metal laid per mile. In addition to the demand for rails, iron was needed for rolling stock, for bridges, fences and stations. In its turn the development of railway communication allowed the further development of the iron industry, more ore and coal could be moved further to furnaces and forges and more and heavier finished products more widely distributed. Export trade in rails grew; as the great period of British railway building ended, French and American railway building reached a peak and both countries were large importers of British rails. Then shipbuilders began to use iron. The first iron ship was launched in 1812; in 1850 9% of new tonnage was iron-built, in 1880 96% of new tonnage was iron-built, but shipping used nothing like the weight of iron that railways did.

Changes in the geographical distribution of the iron-making areas were also great. The Midland district maintained first place throughout, but the area of greatest production moved from Shropshire to Staffordshire. The great increase in the production of South Wales, particularly in the early decades of the century, and of Scotland, particularly in the middle decades, are most striking. Yorkshire and Derbyshire increased their output but at nothing like so rapid a rate.

To understand the expansion of the industry and its distribution in this period it is again of first importance to consider technological changes. Three inventions go far to explain the great expansion of the industry: the invention of the puddling and rolling processes in the manufacture of bar-iron; the invention of hot-blast furnaces; and the invention of the steam engine. The areas that profited particularly by these inventions were areas that not only possessed physical conditions that would allow the new processes and inventions to be used with advantage, this could be said of many regions, but also they were areas at the right stage of economic and social development to take up the new inventions.

The invention of the puddling process allowed coke to be used in the forges to make wrought-iron from pig-iron. The search for a successful method of using coke or coal in the forges was most earnestly pursued in the later eighteenth century, for full use could not be made of the ability to produce pig-iron cheaply and in quantity if the pig-iron had to be converted to malleable iron in charcoal-burning forges. J. Roebuck, the founder of the Carron Ironworks, introduced in 1762 a process using coke that allowed malleable iron to be made from pig-iron, but the result was poor in quality and could not compete with charcoal-made bar-iron, whether home-made or imported. Then independently and almost simultaneously, Peter Onions, a foreman of iron-works, at Merthyr Tydfil in 1783 and Henry Cort, a contractor to the Admiralty, at Gosport in 1784 invented a successful way to use coke to convert coke-smelted pig-iron into good malleable wrought-iron. The cast-iron was broken up, put with clinkers rich in oxides of iron into a reverberatory furnace and heated to melting point. On melting the carbon of the cast iron combined with the oxygen of the clinker, CO_2 was given off, burning as a bluish flame. To help the process the mixture was stirred " puddled " and gradually the pure metal accumulated in a spongy " loop ". The loop was hammered or, better, " rolled " between cylinders to squeeze out any slag. Rolling was an invention of almost as great importance as puddling itself: for it took some twelve hours to hammer out one ton of iron, but in twelve hours, Scrivenor says, fifteen tons of iron could be rolled out.

The hot blast process cheapened still further the production of pig-iron. In 1828 James Neilson working for the Glasgow Gas Company was asked to repair a defect in the furnace for a Muirkirk firm. It occurred to him that furnaces would be much more efficient if the air blown through them were hot, not cold. At the time the idea seemed revolutionary, for it was held that the colder the air blown through the furnace the better the iron: in fact the air was often blown over ice to cool it before entering the furnace. Neilson experimented in using the heat given off by the furnace to heat the air as it was blown along the pipes to the furnace; his experiments were most successful, more and better iron was smelted with much less fuel.

The third invention, the invention of the steam engine, did not belong peculiarly to the iron industry but it was closely linked to it, and full use could not have been made of other inventions without an invention to provide more power, more widely available, than water wheels could produce. In 1782 there is a very interesting letter from James Watt to Matthew Boulton. (James Watt, the Scottish engineer who had taken out his first patent for a steam engine in 1769, foreseeing that the Carron Ironworks were too remote and too small to give full facilities for the development of the invention, had migrated in 1774 to Birmingham and joined Matthew Boulton to found the famous partnership.) The letter, headed 14 Dec. 1782, contains this passage:

"We have had a visit today from a Mr. Cort, of Gosport who says he has a forge there and a grand secret of making

> iron, by which he says he can make double the quantity at the same expense and in the same time as usual. He says he wants some kind of engine, but could not tell what. He wants some of us to call on him and says he had some correspondence with you on the subject. He seems a simple good natured man but not very knowing."

Mr. Cort must have been more knowing than he looked, for he had indeed a "grand secret of making iron" and he realised full use of it needed steam power. The invention of the steam engine came at just the right time for the iron industry: it allowed the new forges to be established near coal and near furnaces without the necessity of finding a site on the already crowded streams; it allowed old furnaces to convert their bellows from water power to steam power, and, perhaps even more important, allowed new furnaces to be built bigger because steam engines could blow bigger bellows to give more powerful draught; and last but not least it allowed iron, and the coal to smelt and forge the iron, to be mined from deeper pits, for there was now an efficient pump to keep dry deep mines sunk beneath the water table.

The distribution of the industry and the changes in distribution, if not wholly explained by these new inventions, certainly cannot be understood without them.

The Midlands in the preceding period added furnaces fuelled with coke to the forges using charcoal established in numbers in the seventeenth century, and took the lead in the production of pig-iron. The Midlands in this period brought their wrought-iron industry up to date, and this involved the resiting of many iron-mills on the coalfields. The most easily mined parts of the Shropshire coal and iron seams were becoming worked out, but Staffordshire coal could be used for coking too and there was plenty of it within easy reach of the surface. The furnaces of the Stour valley, close along the river from Halesowen to Stourport, rapidly increased in number and output. In 1788 Staffordshire had but six furnaces compared with Shropshire's twenty-one, but the position changes now:

Number of Furnaces. 1796-1830

	1796	1806	1823	1830
Shropshire	23	42	35	48
Staffordshire	14	42	84	123

To some extent Staffordshire in this branch of the industry is but catching up, but the great increase in the output of pig-iron reflects too the opportunity that now exists of converting much of it cheaply into wrought-iron in the new forges established on the coalfield. Coal provides not only fuel for use in the forging process but also fuel to make steam to work the hammers. Gradually, as and when water wheels need repair or are obviously hampering production, steam pumps are installed in old riverside mills. Coal can easily be brought downstream and it was found more economic to do this than to run the mill in the old way, or to move it to a new site. The impetus of the going concern is very great; the labour is there, communications and other services have concentrated there, secondary industries may have developed, the social group is in being, a town is in existence; the town now sites the industry rather than vice versa. The function of the river has changed; it no longer supplies power directly, but it allows of the transport of the raw material, coal, that now generates the power. The growth of the industrial towns along the Stour is an interesting example of historical momentum, since the primary condition that localised them passed almost with their birth.

The growth of the Welsh industry in this period was very much helped by the enthusiasm with which the puddling process was taken up by the Welsh iron-masters. The iron industry of South Wales was yet young, there had long been local making for local use, and the disappearance of wood in the Forest of Dean encouraged the migration of forges to the Brecon woods. It was a Welshman who, at the same time as Cort, invented a puddling process. In the Midlands the new process met with some opposition trom established interests, both from those who maintained that there could be no forge-iron like charcoal forge-iron, and from those whose interests were in foundries and in pushing the substitution of cast-iron for forge-iron wherever possible. South Wales, coming into the iron industry in the early nineteenth century, was ready with easily worked clayband ores alongside easily mined coal to take advantage of the newest methods unhampered by preconceived ideas and thereby profited.

Scotland came to the fore a little later for a very similar reason, the readiness to profit by a new way of doing things. It was a Scotsman who invented the hot blast process; a process particularly well suited to the Scottish conditions. It allowed coal to be burnt in the furnace instead of coke, and the Scottish coals did not coke as well as those of Staffordshire and South Wales. Not only was the difficulty and cost of coking coal avoided, but also the amount of fuel used was much reduced; it was claimed that two and a half tons of fuel was saved on every ton of pig-iron produced. This made it worth while to use the local "blackband" ores. Hitherto their low iron content made it hardly economic to smelt them. Now the impurity, "the black" in the band, which was largely coal, was an asset and these ores could be burnt almost without fuel. The new process allowed very cheap production and since the coal and iron fields were near the magnificent navigable water of the Firth of Clyde and its rivers, the remote situation was no disadvantage. Transport by sea was cheap: Scottish pig-iron could compete in cost and quality in all markets. Between 1830 and 1850 about a quarter of British pig-iron was produced in western Scotland and furnaces in south Staffordshire were going out of blast, and Wales lost her dominant position in spite of a considerable import of high quality haematite ore from Cumberland. The dominance of Scotland was greatest about 1865 but it was short-lived; her blackband ores were limited in quantity and by 1870 north-eastern England surpassed Scotland in percentage ouput. This brings us to the period that immediately precedes the modern age.

The Age of Steel

Between 1860 and 1913 the steady increase in the make and use of steel and the shift of the chief iron-making centres to coastal coalfields are the maior new features.

The figures given in Table IV show the general trend of industrial output.

TABLE IV

Pig-iron Manufactured in 1865, 1884 and 1913. (Output given in million tons.)

District	1865	1884	1913
Wales and Monmouth	.91	.88	.89
Shropshire Staffordshire	.11 .89	.55 .57	} .85
Derby Yorks	.18 .12	.35 .24	— .50 Yorks & Lancs
Derby, Leics, Notts, Northants	—	—	1.17
Furness, Cumberland	.31	1.50	1.16
Cleveland, Northumb, Durham	1.01	2.50	3.87
Scotland	1.16	.98	1.37
Total	4.69	7.57	9.81

Figures for 1865 and 1884 given in Sir J. Lowthian Bell, *The Iron Trade of the United Kingdom* (1886); for 1913 given in Annual Volume of Statistics published by National Federation of Iron and Steel Manufacturers.

There is a steady rise in total make, though not as spectacular a one as in the preceding period; the products of European and American iron industries are beginning to oust British iron and iron goods from their dominant position in world markets.

There is in this period a further change in the geographical distribution of the industry. The north-east coast, Teeside, and to a less extent the north-west coast. Whitehaven and Barrow-in-Furness, show a remarkable growth; Scotland and Wales decline from their dominant position, and the Midlands come to play a relatively minor role. The change in geographical distribution is brought about once more by first a change in technological processes, and secondly a change in the value of geographical resources.

In the Midlands, particularly in Shropshire and south Staffordshire, the best seams of the clayband ores of the Coal Measures that had been continuously worked since medieval times were becoming exhausted. There was still ore in the region, but only the thin irregular seams at considerable depth were left, and in competition with Jurassic ores mined at or near the surface in Yorkshire, Lincolnshire and Northamptonshire, and with ores imported from Spain and Sweden, it was no longer economic to work them. The best coal seams were also worked out. The favouring conditions of ore and fuel and water-power that had done so much to make " the Black Country " in name and fame had passed ; skill alone remained ; the industrial role of the Midlands reverted to what it had been in the medieval and Tudor and Stuart period, it worked pig-iron imported from elsewhere but on a very different scale. The manufacture of steel and of an immense variety of iron goods took the place of the forging of bar-iron and the manufacture of nails. In Scotland and in the north-eastern region of the South Wales coalfield, the easily worked coal and iron beds in close proximity on which the iron industries of the Coal Age depended were also rapidly becoming worked out. The ore and coal left were more difficult and more costly to work and in spite of the great inertia created by its heavy equipment and large labour supply the iron industry gradually moved away, leaving its old haunts vastly changed as a result of its growth and much distressed by its decay.

A technological revolution demanding new conditions added its effects to changes in local resources of ore and coal. In August 1856 Henry Bessemer read a paper to the British Association with the arresting title, " The Manufacture of Iron without Fuel ". His claim was to have found a method of making steel without the immense consumption of fuel which raised the cost of this form of iron to a level which prohibited its use, in spite of its advantages, except for very special purposes. Pig-iron was not pure, it contained a small amount of phosphorus and often other elements and a sensible amount of carbon. Pig-iron melted and run into moulds directly—cast-iron—contained about 4 to 5 % of carbon; pig-iron heated and hammered and refined in chafery and finery forges—bar-iron or wrought-iron—contained only about 0.3 % carbon. Cast-iron was hard but brittle, bar-iron was malleable and easily worked and had much more tensile strength than cast-iron but was not very hard. For some purposes, to make for example knives which would keep a cutting edge and axles and pivots that would show little wear in spite of friction, an iron was needed which had the hardness of cast-iron but the ductility and tensile strength of bar-iron. Steel, by definition an iron containing 0.1 % to 1.5 % of carbon, is very hard if cooled suddenly and has greater strength and elasticity than wrought-iron. Different steels vary in carbon content and, also, especially in recent times when alloyed with other metals, in hardness or other properties.

Steel had long been made in small quantities : the cutlers had made "blister" steel by breaking up bar-iron of finest quality, binding it up into bundles with faggots of charcoal and heating the bundles for several days to a very high temperature in a charcoal fire. The iron thus heated increased its carbon content. The "blister" steel was then broken into fragments, mixed with sand and reheated in an air furnace and then hammered to give a malleable metal, "shear" steel. The cost of steel thus made was very great because of the immense amount of fuel needed to maintain the high temperatures so long. In 1742 Benjamin Huntsman, a Sheffield clock maker, found a way to make better and more uniform steel—a cast steel—by melting small particles of blister steel and shear steel in a crucible, maintaining it at a high temperature for five hours and adding a secret flux to free the metal from particles of silicates and other impurities. The cost was still very high ; 1 ton of steel rails in 1855 cost £75. No wonder then that Bessemer's claim created a stir. His method was to run liquid iron from the furnace into a "converter" and to blow a powerful blast of air through the molten metal. The carbon in the iron combined with the oxygen in the air and was given off as CO_2. As much carbon as was wished was then added to the iron, usually in the form of spiegel, an alloy of iron and manganese with a carbon content of about 0.3 %. The method worked extremely well with ores which contained little phosphorus. Unfortunately all the British ores of the coal measures, both blackband and clayband, and the ores of the Jurassic sandstones and limestones were phosphoric ores; only the haematites (Fe_2O_3) of Cumberland and Furness, and imported Spanish ores could be used. But with the right ores steel could now be produced cheaply and in quantity. Steel was still further cheapened by the invention in 1861 by the Siemens brothers of a process later modified by Martin. Cast-iron, scrap-iron, and iron oxide in calculated quantities were heated in an open hearth in a furnace. The oxygen in the iron oxide burns out the unwanted carbon in the cast-iron, leaving a sufficient carbon content in the mixture to give steel. The great advantage of the process was that it used scrap-iron which further cheapened manufacture, but for a time more steel was made by the Bessemer process which had the advantage of an early start. The cost of a ton of steel rails had been reduced by 1864 to £17 10s. A method of using the native ores to make steel was discovered in the next decade. By 1879 a patent had been taken out by Thomas and Gilchrist whereby the Bessemer converter or the open-hearth was lined with dolomite (magnesian lime) which combined with the phosphorus in a phosphoric ore and produced good steel. This was known as the "basic" process, and the steel so made known as basic steel, in distinction to the Bessemer and Siemens-Martins processes which, because the converter or hearth was lined with silica brick, was known as the "acid" process which produced "acid" steel. The basic process was not immediately widely taken up, the acid process was by 1879 well established; it would have been costly to change over, and the royalties on the basic process were also heavy. The sequence in which the discoveries were made and the relative patent costs had a major effect on the geography of the industry. By the 1880's, however, all three processes were in use, but it was not until the European War of 1914-18 that the basic process made great headway against the acid process as the following figures show.

TABLE V

Production of Steel (in millions of tons).

	1870	1878	1880	1883	1885	1900	1905	1913	1923
Acid Steel									
Bessemer	.23	.8	1.	1.5	1.2	1.6	2.0	1.6	.4
Open Hearth	.15	.17	.25	.46	.6	2.4	2.4	3.2	2.4
Basic Steel	—	—	?	.12	.14	.8	1.2	2.2	3.4
Wrought-iron			2.7	2.7	2.2	1.1	.9	1.1	.3

The make of wrought-iron does not fall in competition with steel until about 1885, because home and world demand for all kinds of iron was increasing. Great impetus was given to steel production when first the railway engineers and then the shipwrights went over from wrought-iron to steel. In 1877 the N.E.R. ceased to buy iron rails and began to replace iron rails as they wore out with steel ones. It was nearly another decade before steel replaced iron for ship-plates and anglebars, steel at first was not sufficiently cheap or sufficiently uniform in quality to satisfy the shipbuilding industry.

The effects of the new processes of steel manufacture, and of the sequence in which they were introduced, on the geography of the industry can now be considered.

The north-west of England had made a brief appearance in the middle of the eighteenth century as a centre of iron manufacture, but once its woods were depleted the iron industry declined. Coal was not easily obtained in the area, and the ore that was mined was exported to be smelted elsewhere, particularly to South Wales. In 1852 the amount of iron made in the area was negligible. The introduction of the Bessemer process entirely changed the scene. The ores of Furness and Cumberland were the only non-phosphoric ores Britain possessed, and this alone ensured attention being given to them. The seaboard situation, the nearness by rail to the Durham coalfield, allowed the import of good coking coal and it was not long before a way was found to use successfully the local coal too. In 1865 Cumberland and Lancashire once more appear among the important areas of iron production, and by 1875 the local industry consumes the whole output of local ore; export of ore to Wales, Scotland and the Midlands has ceased. By 1884, it is the second largest producer, only surpassed by the north-eastern area. Whitehaven and Barrow grow from small unimportant villages into "planned" towns laid out with rectangular streets almost within a decade. But by 1913 not only the relative importance of the area but also the total output of the iron-works was decreasing. The supply of easily accessible local ore was limited; the industries based on iron manufacture in the region were small, they were not sufficient to pay for the import of raw material to keep them going once local resources gave out.

Conditions in the north-east were quite different. The Tees-Tyne region too had little importance before the nineteenth century but, about the middle of the century and before the invention of the new processes to make steel, it was coming to the fore in the production of wrought-iron. About 1850 the Jurassic ores of the upland belt sweeping from the Cotswolds through Northamptonshire and Lincolnshire, to the Cleveland Hills were once more receiving attention. In fact even the knowledge of their use in the Middle Ages had been lost, and their existence was now hailed as a discovery. The ore was lean; it had a content of about 30-33 %, a richness much the same as the clayband ores, but much less than the haematite ores of the north-west. Unlike the clayband ores, however, these Jurassic ores did not occur associated with coal and thus once the wood in the immediate vicinity of a quarry or mine was used up the iron industry based upon them declined. By the middle of the nineteenth century, however, it was clear that the Coal Measure ores which could be mined would not last long, and thus these Jurassic seams near the surface would obviously become important if fuel could be brought to them. The discovery that the ore outcropped to the north of the Cleveland hills, within about five miles of the Tees estuary, where an iron industry existed already, based upon Durham coal and imported Scottish iron, laid the foundation of a great industrial development along the lower Tees. The industry centred in Middlesbrough profited by the fact that it was growing during a period of great technological advances. The furnaces were built to take advantage of new methods; waste gases could be used to raise steam and to heat the blast; furnaces were built larger because, with a hot blast, a higher temperature could be raised in the bigger furnace and thus each unit smelted a much bigger volume of ore. A great saving in fuel and labour was effected, and Lowthian Bell remarks that "before long all the small furnaces on the banks of the Tees will be demolished".

The seaboard situation was a great asset in the distribution of the finished products. Rails were heavy and cumbersome and could most easily be transported even to other parts of England by sea, and the shipbuilding industry of the north-east coast was fed more and more by locally made iron.

The region was peculiarly well suited geographically and economically to take up the new inventions. The first Bessemer converters were erected in Sheffield, but the local iron ore was phosphoric and unsuitable. Spanish ore had to be imported. The Don was not navigable, ore could be brought so far up the Trent and the Idle but there was at the end a land haul to Sheffield. It was obviously better to import not ore but pig-iron. But the import of pig-iron meant that the advantage of running molten iron direct from furnace to converter was lost, so that it was an obvious step for Tees-side to set up converters to make steel beside the furnaces more conveniently placed to use imported ore. The north-east coast region built its immense steel industry on abundant cheap good quality coking coal, on a seaboard situation which allowed the cheap import of Spanish and later Algerian non-phosphoric ores and the export of finished goods, on the experience of an established iron industry. Soon it was cheaper to make steel rails from Spanish ore in Middlesbrough than to make wrought-iron rails from local Cleveland ore. As the great iron and steel industry of Tees-side grew, it fed in turn other industries: the shipbuilding yards on the navigable waters of Tyne, Wear and Tees constructed 50% of British built ships; the marine engineering shops, the locomotive shops, the constructional steelwork industries, all grew using the twin opportunities offered by abundant raw materials on the spot, and deep-water communications for their heavy products to home and overseas markets. The import of foreign ore was economic, in part because the ships bringing it in could take out the coal and manufactured iron goods which were in great demand in the ore-producing areas.

Scotland and South Wales, though they decline relatively in the Steel Age, maintain an important industry largely because they too have a coastal situation, and the invention of steel-making came at just the right moment for them. Along the Clyde and at Carron foreign ore could be cheaply imported to make steel to take the place of the clayband and blackband ores, now becoming too scarce to keep going on the old scale the making of cast- and wrought-iron. Alongside the iron and steel works grew up the heavy industries, and among them the shipbuilding of Clydebank. In South Wales change-over to new methods brought about an internal change in the location of industry. The outcrop of clayband ores near rich and easily worked coalfields had encouraged smelting and puddling in the upper parts of the little valleys tributary to the Usk, Ebbw and Taff, and the growth of a long line of mining and industrial villages from Bryn Mawr through Ebbw Vale and Tredegar to Merthyr Tydfil. This industrial area was a long way from tide-water; some ore was imported, some steel works were set going, for there was in the district a great reserve of skilled labour, and a great amount of capital sunk in the expensive machinery of the mines and furnaces. But gradually the ports at which the ore arrived, either from Furness or from Spain, Llanelly, Swansea, Briton Ferry and Port Talbot set up steel works; coal could be imported too. Here associated with steel making grew too, the tinplate industry; the tin was imported first from Cornwall, then from Malaya. In contrast to Tyneside, Barrow and Clydebank, shipbuilding has never been of much importance in South Wales.

The accident that the acid process of making steel was invented before the basic has done much to cause the coastal location of the industry, for it was the method of manufacture that created the demand for foreign ores. In Britain the acid steel processes held their own until the twentieth century. The local Jurassic ores were lean compared with foreign ores, and, except in Cleveland, outcropped far from coalfields. Only in Middlesbrough was much use made early of the basic process. Coming to the fore at the time of the invention, progressive in its ways because all its capital was not already tied up in older concerns, Middlesbrough built basic hearths and smelted Cleveland ore into good steel. Foreigners came from far and near to learn at Middlesbrough and the Germans in particular learnt to some purpose. The German coalfields lay inland, it was not cheap and easy to import foreign ore to them. Lorraine ore was phosphoric; the great industries that developed by marrying Lorraine ores to Saar and Ruhr coal to make steel by the Thomas and Gilchrist basic processes became the mainstay of the economy of western Germany.

The modern geographical pattern of the iron industry begins to emerge in the second decade of the twentieth century. The conditions brought about by the European War, 1914-18, threatened to cut the supplies of ore from abroad, shipping was scarce, submarines took their toll of what ships could be spared to carry iron, and the Jurassic ores of the eastern Midlands for the first time since the Middle Ages became of major importance again. They outcrop near the surface, open cast mining is cheap though messy, and railway transport allows ore to be taken to coal or coal to ore for smelting. Since with up-to-date processes and low quality ore it takes less coal than ore to make a ton of pig-iron, the tendency is for smelting to move to the iron fields: the hold of the coalfields on the smelting industry is loosening now as the hold of the forests loosened over two centuries ago. The pig-iron so produced needs much coal to make it into steel, therefore the coalfields tend to attract still the steel industry, and in particular those coalfields best placed to send finished steel to the biggest markets. In 1850-70 Britain produced one-half of the world output of pig-iron and easily led the way in the production of steel; in 1913 she made only one eighth of the pig-iron and one-tenth of the steel produced in the world and she launched only 1.9 million out of a total of 3.3

million tons of shipping. Though her export trade increased temporarily after the 1914-18 war before the ravished Continental industrial areas recovered, it never regained its former dominance. Thus to supply the home market the central position of the Midlands has much to recommend it in preference to the peripheral positions of Clydeside, Tees-side and South Wales.

The geography of the industries of most regions will afford examples, as striking as in Britain, of the continuous change in scene and the continuous revaluation of geographical conditions that time brings about. The hegemony of Byzantine industry passes with changes in the balance of political power and consequent changes in trade routes. The international repute of Tuscan silk, Venetian glass, Cordovan leather, Toledo steel, Sèvres china, Dutch atlases calls to mind later industrial cities, which in turn were eclipsed when coal became king. Even the New World with its relatively short industrial history shows great changes in industrial geography. In the colonial period New England led in the industrial field. New England's industries in themselves show interesting changes in location. At first, since her cloth, iron and ships sold in a much wider market than her own, her industries were established in coastal towns for easy export. When machinery which needed considerable water power became a *sine qua non*, industries migrated to the Fall Line. Finally they came back to the coast again, drawn by the need to import raw materials and coal for power, and to maintain the closest possible links with home and foreign markets. But the tale of New England is the tale of some parts of England over again. New England used her resources early, but in time she used them up; she had experience but she was hampered as well as helped by tradition. Coal and iron occurred together near Pittsburgh; Pittsburgh iron had a low phosphorus content, it made good steel by the acid processes. Chicago, like New England in later days, had to import both iron and coal; Chicago had no local tradition of manufacture or reserve of skilled labour, she had to import teachers too; but Chicago had an immense market at her door in the West, a market for rails, for locomotives and rolling stock, for constructional steel for bridges and buildings, for mining and agricultural machinery. New England's iron and steel industries fell behind in competition with Pittsburgh and Chicago. The raw cotton produced in the South and big market in the warm lands for cotton cloth were conditions which favoured textile manufactures in the South. The exhaustion of the soil which arose from plantation agriculture, the soil erosion that followed unskilled farming of abandoned cotton lands, and the freeing of Negro slaves helped to create a redundant agricultural labour supply, Negro and white, which could be profitably set to work in cotton mills. The factories of Georgia and Alabama came to rival those of Boston.

Industrial geography wherever studied thus presents a constantly changing scene; old industries use new materials and new sources of power to make by new processes better and more varied goods; new industries arise and replace old; new demands are made by old customers, and new customers increase the demand for well-known products. The area in which by chance or by design an industry has become established long ago may be able to provide a favourable setting for each new process as it is introduced, may be able to meet each new circumstance as it appears. It may be as well placed for supplies of new fuel as of old, it may be able to import economically raw materials which can no longer be found locally, if old markets fail it may be that easy communication and cheap transport can be established with new. Some areas seem to go on from strength to strength, able to be adapted to fulfil the needs of each successive generation. In these areas the geography of industry is exceedingly complex; just because the pattern is so old it is difficult to discover the conditions that favoured its origin, and to appreciate all the circumstances that have fostered its evolution. The threads of time interlocked to make the design in space, are extremely difficult to disentangle. However, it is in just these areas that it is most necessary for the geographer to study past conditions and to remember the power of historical momentum and the strength of geographical inertia. In contrast, in other areas the circumstances which once fostered an industry may have been so specialised that they are not sufficiently elastic to allow advantage to be taken of later technological improvements or inventions, or of changing supplies or varying demands. An old industry may linger on as a relic or may die out to be replaced by a new one. It may even give way to a purely agricultural economy, and the one-time industrial towns and villages will then serve as the market and social centres of neighbouring farms, helped out perhaps by a tourist trade fostered by the architectural heritage left by the wealth of masters and merchants in preceding generations. Here too the past is the key to the present.

The changing geography of industries in any one area is bound up with changes in other areas which complement or compete with them, and with changes in the world market. Flourishing industry depends upon good communications; communications that are "good" in one age may be ill-suited to the traffic of the next age, thus the changing geography of communications and transport is also a theme of real importance to the geographer.

Map Exercises

Suggestions of maps or diagrams that the reader may make of his own area to illustrate this chapter.

1. A map of the industries remarked on by, e.g. Leland, Celia Fiennes, Defoe.
2. A map of water mills showing their uses.
3. A map of mines giving the dates at which they are first recorded as worked.
4. A map of present factories giving the date on which they were opened.

CHAPTER VIII

THE CHANGING GEOGRAPHY OF COMMUNICATIONS AND TRANSPORT

THERE may be natural resources much prized by man that lie unused because they are too difficult to reach. "Difficult" is a relative term, here it means costs more in time or effort, and therefore probably but not necessarily also in money, than man is prepared to pay. If man wants to go to any place enough he will find means of getting there, whether on foot or in the latest aeroplane, and if the stimulus is great enough he will find ways of transporting what he wishes to transport to or from the area, whether on the backs of women or llamas, or in the trucks of lorries or the holds of ships.

It is often said the the sum of the effects of the physical and biological conditions of a region can be seen in the geography of the natural vegetation ; it might equally well be said that the measure of the value of the natural resources and of the intensity of economic, political, and social activities based on them can be seen in the geography of communications and transport. The geography of transport is therefore suitably the last single theme as the geography of peoples was suitably the first to consider. Paths and roads, navigable rivers and canals, and railways are in themselves conspicuous geographical features ; but their greatest importance lies not in this but in the function they perform. The pattern of settlement, of agriculture and industry, is closely related to the pattern of communications, natural and man made.

The study of the geography of transport certainly demands study in past periods as well as today. In some areas and in some ways the geographical pattern of communications has great stability and old patterns have exerted striking long-term effects. In other areas and in other ways the pattern changes rapidly, sometimes dramatically, and thus affords an excellent chance to see *how* geographical changes take place and to observe the consequences. The lines of communication increase in number, making as time goes on a closer and closer network, and the means of communication change too as new inventions make practicable that which once seemed the wildest dream. Wheels succeed feet; the steam engine, the motor engine, the jet engine—and the atomic engine?—succeed ox and horse power; telephone, telegraph, and wireless succeed fire and drum, the runner and the penny post. Geographical distance may not change measured in miles, but it certainly changes enormously measured in time and money. Since geography in essence is the study of space relationships, distances are of first importance and a revaluation of distance entails a revaluation of almost all, singly and in sum, of the features, factors and agents that contribute their mite to the character of the place.

The geography of transport needs examination in any area at whatever period, and for whatever reason, it is desired to make a study, but it would seem that to realise the nature of the communications may be particularly important in dealing with three periods. The first is the phase of the settlement: the groups who colonised, the routes of the colonisers, when and where they took up land, is always closely related to the opportunities of transport. The second, in regions early settled, the British Isles and Europe, for example, is the long period which followed the first expansion of the settlement and preceded the railway age. What might be called the "medieval" communication system is then at its height, a system that depended much on careful and intricate organisation but little on mechanical power. The opportunities and limitations of this system are too frequently unrealised, or at least not sufficiently considered, by geographers discussing the problems of, for instance, the location of a town or an industry, where the communications existing not only at the time of birth but during the period of early growth matter so much. The third is the early days of steam power, when railways and steamships had largely replaced animals and sails for heavy and long distance traffic, but before the motor car, the motor ship and the aeroplane take pride of place. This in many areas is a period that inaugurates immense changes, fundamental to the understanding of the present geography.

Too much is often said to be the effect of the " lack of transport there ", without giving any real thought to what in fact is known of the communications and transport *there then*. This chapter will try to establish the elementary facts for earlier periods and will only briefly and indirectly refer to their effects. The general principles of cause and effect as exemplified in the geography of transport apply at all periods, and for a discussion of these readers are referred to the appropriate chapter in another volume of this series—W. S. Thatcher, *Economic Geography*. In Britain the main features of the geography of the land outside the highly industrialised areas were drawn and sculpted before the railway age, but those of the industrial areas owe their character very largely *to* the railways and all that they meant.

The Pattern of Communications in Britain

The most characteristic features of the geography of communications in Britain today might be said to be first the dominance of road and rail traffic within the area and the dependence on ships for the greater part of the transport of goods to and from the rest of the world, and secondly the nodality of London in both land and sea communication networks, with a few other centres of much less importance.

Internally all the main railways and most of the main roads radiate from London. A secondary network within these centres on each of the regional capitals, Birmingham, Manchester, Leeds and Glasgow, and to a lesser extent on Bristol, Liverpool, York, Carlisle and Edinburgh. The main A. roads, with a few exceptions in mountain areas, are relatively straight, the secondary roads (B. roads, and still more C. roads) in many districts twist and turn in a surprising way, often for no cause visible now. The closeness of the road and railway network is in general a reflection of the density of population but sometimes a maze of little roads serves a remote area. The pattern serves modern needs but it has grown slowly ; it contains within it the skeleton of earlier days and new " spines ", " trunks " and " limbs " are usually grafted in. Only occasionally a completely new line is driven straight through, breaking up the old design. Authorities widen here, cut off a corner there, make a by-pass to avoid congested town or narrow village streets, and so old roads live on adapted to modern needs. The grid-iron road pattern of new towns and suburbs shows up alien and strange in Britain. But if traffic deserts a road the thoroughfare itself soon disappears. The footpath becomes ploughed up, the bridle track overgrown, the road an archaeological "find" seen perhaps on an air photograph only. The right of way is lost; highways unused die. In order to attempt to understand the present communication system and to appreciate the location of almost all else in Britain, the old bits of the present network must be identified and the discarded bits now overgrown recognised as a preliminary step.

Local Study: Sources

The best way to realise the stability on the one hand and the essential changes on the other in the geography of communications, and to appreciate its interaction with other factors in the choice of sites for settlement and in the growth of agriculture and industry is for the reader to consider the communications of the area about which he daily travels. Let him pause to think what it was like to get about in it before there were trains and buses, tubes and motor cars. A good beginning is to examine an early edition of the one inch Ordnance Survey sheet of the area; in many places one more or less contemporary with the first railway lines will be found. If an Ordnance Survey map of the right date cannot be found, then the local map from one of the numerous atlases of county maps published about the middle of the nineteenth century may be consulted instead. A history of the railway that serves the local area and the files of the local newspaper when railway schemes were afoot will provide further information. From the accounts of the often acrimonious early meetings at which storms of protest greeted the proposals of the "promoters", through all the long discussions, to the final reports of the jollifications and free rides that celebrated the opening of the line, much understanding of the local contemporary geography will be gained. A map made to show the first railway lines and stations with their dates is often most revealing of the communication needs of the area as envisaged then and may suggest the cause of later changes. The contemporary roads may be added to see how far they feed and how far they duplicate the railway lines. If a contemporary timetable can be found, a diagram may be made showing the trains per day and usefully compared with a similar diagram made for the present day.

The period of the building of the canals may next be considered and a map made to illustrate it. The canal-building era may be before the foundation of the local newspaper but the dates of the Canal Acts and for some of the opening of the canals may be found in H. R. de Salis, *A Chronology of Inland Navigation*. The length, width and draught of the boats taken, and the position of locks and tunnels may also be shown.[1] The portions of rivers navigable at that date should be marked too; stretches navigable now were probably navigable then except where recent 'improvements' have been made. The great period of 'improvements to navigation' precedes the Canal Age. It is well to remember that higher reaches and smaller streams than it is worth while to use now may have been navigated by small boats in the eighteenth century. Contemporary roads to add to the map will be found in Owen's *New Book of the Roads* (2nd edition 1779) or Paterson's *New and Accurate Description of all the Direct and Cross Roads in Great Britain* (1st edition 1771), or Cary's *New Itinerary* (1st edition 1798).

The roads of a century earlier still can be mapped from John Ogilby's *Britannia* (1675). Ogilby carried out the first measured survey of the English roads, and incidentally brought into common use the statute mile of 1,760 yards. His road book was reissued again and again; it was "revised and improved" by several editors, in particular, John Senex and Emanuel Bowen. If one of these "revisions" is used, it is well to remember that the roads of the area depicted may not be the roads as they were in 1675 but as the editor believed them to be at the time of the revision. Local guides of the eighteenth century often give

[1] Bradshaw's *Canals and Navigable Rivers of England and Wales*.

in an appendix the regular service of the stage-coaches and waggons, citing the days of travel, the places of call and the time and cost of the journey. The little guide to Cambridge entitled *A Concise and Accurate Description of the University, Town and the County of Cambridge* of 1763, for example, tells of stage coaches that ran daily to London, Lynn, and Ipswich; twice weekly to Wisbech, and weekly to Birmingham. It gives, too, waggon services to London, to most of the market towns of East Anglia, to Wakefield, to Sheffield and to Manchester. The reader can probably find similar guides to one or more towns within his area and so make a diagram of the regular traffic along the local roads and of the links with more distant places.

For periods earlier than the seventeenth century anything approaching a complete picture of the communications of an area cannot be made, but some clues to the system may be gained. It is suggestive to mark on an outline survey sheet, for example, the portions of Roman roads and prehistoric track-ways that have been identified in the area, and to add the medieval bridges, like those of Wakefield and St. Ives with their way-farers' chapels, or guarded by towers as the famous London Bridge, or merely recognisable by their masonry, the shape of their arches, or their narrow width. Those in ruin as well as those in use should be plotted. There are, too, the little pack-horse bridges, so narrow that certainly no cart could cross, but with parapets low enough for the packs of laden animals to clear them. These gave safe crossing over streams too deep and swift to ford. They are more common in the north and west hill country than in southern and eastern plains where fording was easier. Many of the pack-horse bridges are seven-teenth and eighteenth century in date, but many are earlier and some of these can be approximately dated. Fords can be mapped too, but unless there is documentary evidence they cannot be dated. Place names that contain elements that refer to streets and crossings may further help to establish a route early fre-quented. *Street* in all its various spellings—*streat*, *strete*, *stret*, *strad*, *strat*—means a paved road. The places with 'street' in their names, unless the name is modern, all lie on or near a paved causeway, and since the Romans made the great majority of roads *built* before the nineteenth century most of these names lie on or near a Roman road. All the numerous Stratfords are on Roman roads, so are the Streets and most of the Strattons and Strettons. Streatham (Surrey) is on Stane St., Strefford (Salop) is on Watling St., the Stratfields (Berks, Hants) on the Roman road from London to Silchester. It is perhaps wise to repeat here the warning that all place names, however obvious their meaning may seem, must be looked up before deductions are made from them. There is the Welsh word *ystrad*, a valley, that appears too as *street* or *stret* to trap the unwary; even *steorc*, a young bull, may be transformed into a street by the careless—Stretton (Cheshire) which is Strecton in 1287 may have this derivation. The *-bridge* names, Bridgnorth and Bridgwater, Cambridge, Tonbridge, Boroughbridge and Coatbridge, for example, suggest a crossing of some importance, otherwise the expense of construction and upkeep of the bridge would not have been worth while. It must however be remembered that towns at important bridges may bear no sign of this in their names, and the existing bridge may be of obviously modern construction. Records of medieval tolls may establish at least a limiting date for the existence of a bridge. It is probably safe to assume that most rivers in their upper and middle reaches could be crossed by bridge or ford at the bigger towns on their banks. *Ford* names are very numerous: Oxford, the ox-ford, Bradford, the broad ford, Stamford, the paved ford, Cranford and Carnforth, the cranes' ford and many more come at once to mind. A few crossings were too wide to be bridged or forded; there a ferry was needed and some place names record the existence of one: North and South Ferriby, the *bys* at each end of the ferry across the Humber, North and South Queensferry lie on either side of the Forth, and Broughty Ferry is on the Tay. It is interesting and instructive and certainly amusing to try to visualise with the help of the place names the early routes that linked the odd lengths of Roman road and of prehistoric trackways that archaeological finds can establish (Fig. 14). No definite solution will be obtained, for the routes off the main roads and between the crossings were rarely direct; they wound in and out to pick up villages, to avoid wet ground and close wood, and to seek high dry ways with a view. But the exercise will nevertheless serve to imprint on the mind the part old routes may have played in the location of present places.

In a coastal area the dates at which quays and piers were built should be sought, local taxation lists examined for records of the port, and customs accounts studied for information about the goods exported and imported and the places to and from which they were consigned.

Local studies pieced together will eventually provide evidence of the geography of transport of a wider area, and a local study once made should be considered, in so far as information allows, in its regional setting as a part of the communication system of Britain as a whole, and indeed of Europe and ultimately of the world. Wool and cloth from English farms and villages reach Italy, and spices and silks from the Far East reach English country markets, by slow and devious but closely integrated land, river and sea routes centuries before navigators could find their way across the open oceans.

The Communications of the Period of the Settlement

Though in general and over the centuries it is certain that the early settlers colonised widely, particular individuals, with the exception of soldiers and itinerant traders, probably migrated only short distances in primitive boats along the nearest navigable stream, or on foot along the most open ground. The day-to-day movement of most of them was along pathways and cart-tracks in and about the village and its fields and occasionally 'over the hill' or 'down the valley' to the next village. If the little dusty or muddy alleys and open spaces of the villages and the more or less distinct paths that link them to one another of a Balkan country or of parts of South Italy or Spain are recalled, the communication system which affected the siting of the villages and farms and in turn grew up to serve them will be better visualised. In fact even today in the remoter districts of Wales, Scotland and Ireland, there are many farms and even a few villages which cannot be reached by car. Some of these trackways undoubtedly survive incorporated in the existing road system, though all trace of their age and original character is for the most part lost. Here and there in areas of late colonisation where the three-field system never became well established the modern road system is extremely complex; narrow roads wind in and out with many sharp rectangular bends, still preserving the pattern of the tracks between and around the edge of little clearings, the "fields" and "leys" early enclosed by inviolate boundaries.

How visible, how well worn were the so-called prehistoric trackways along the Downs, or the age-old routes swinging in a wide curve from south-west to north-east along the scarp edge of the chalk (the Icknield Way) or of the Jurassic limestone-sandstone outcrop (the Fosse Way) before the Roman invasion is debatable. Broad, perhaps shifting, tracks can still be seen beneath the fields in many places along these routes but they may have been worn any time before the enclosures, for these highways running with the grain of the country remained in use throughout the centuries. Modern roads coincident or parallel with them are but their most recent version.

The roads built by the Romans which receive so much attention may well have had less effect on settlement and work than the old meandering paths and ancient highways worn by the villagers. The Roman roads, however, are the first integrated system of routes radiating from London, giving this lowest crossing of the Thames a nodality which it is very probable was never wholly lost (Fig. 50.) Many first-class roads today follow, in part in fascinating detail, the Roman ways. Many examples can be found on Ordnance Survey maps or on the road maps of motoring atlases. The Great North Road (A.1) from London to Edinburgh, for instance, picks up again and again the Roman road from London to the Wall, later called Ermine Street. Ermine Street branched just north of Lincoln; one branch ran north along Lincoln Edge to the ferry across the Humber, the other skirting the southern edge of the marshes of the Trent and Idle valleys ran through Littleborough to Doncaster and thence north through Castleford and Catterick along the eastern edge of the Pennines to the frontier wall. The modern Great North Road (A.1) deviates from the Roman road to the westward south of Grantham but picks up the western branch of the Roman road again at Bawtry. The Roman link from Bawtry through Littleborough to Ermine Street is still followed by a secondary road. The modern roads along the western edge of the Pennines do not coincide very closely with the Roman ones, the recent development of the Lancashire manufacturing towns has broken up the old pattern, but the

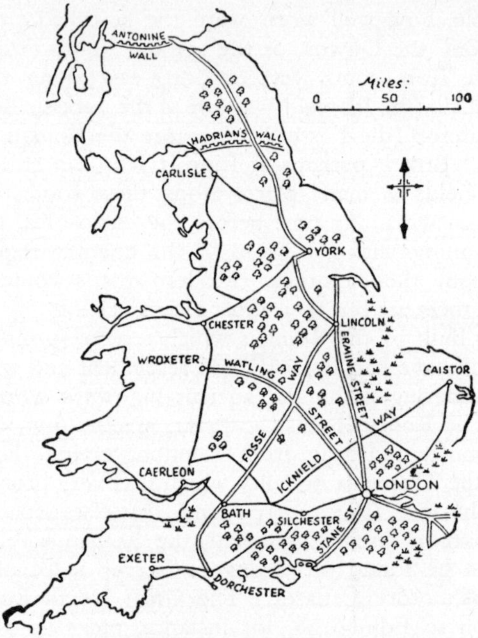

Fig. 50. The Roman Roads of Britain.

links across the Pennines still follow in parts Roman routes. The main road London to Holyhead (A.5) is in its eastern section the Roman road to the Welsh frontier, Watling Street, and the London to Dover road (A.3) through Rochester, Crayford and Canterbury is the Roman route to the port.

How is it that the geographical pattern of roads built in the first and second centuries A.D. shows still in the twentieth century?

The Roman roads were built, it is thought, by army engineers and army labour to fulfil the needs of the administration, military and civil, for quick direct communication between the Continental ports and the centres of civil administration in the south and east, and the fortresses and defence lines of the furthest frontiers in the north and west. They were driven straight across the country, regardless of the cost of construction or the needs of local traffic. The routes seem to have been aligned from hill to hill, forest cut along the track, foundations carefully laid and even causeways built across short stretches of marsh. Large fen or marshy areas did cause deviation. The modern first-class roads fulfil much the same function; they provide for quick movement by direct routes from London to the main provincial centres. This similarity accounts largely perhaps for the similarity of the radial pattern between the Roman and the modern systems, but it will not account for actual coincidence of route.

Coincidence of route does occur too often to be fortuitous, though it is not complete. There is least coincidence in detail between Roman and modern roads where physical conditions offer a wide choice of equally good routes, as over much of the western Downlands, and there is also little where there has been much recent change in population and settlement, for example around London and in industrial Lancashire. In many parts of the country the general line of the Roman road and of the modern road is the same but the modern road has shifted downhill; probably dry ground determined the line of the Roman road, valley villages grown since Roman times the line of the modern one. This shift can be seen for example in Wharfedale, where the modern road takes the valley and the Roman road swept across the moors, and in Berkshire, where the Great West Road follows the Thames and the Roman road through Staines to Silchester kept well up the southern slope of the valley side. The modern road runs along the floor of the Lea valley to Ware, the Roman road lay west of it higher up the valley side; the road today from the Stour valley comes into Cambridge along the valley of the Granta, the Roman road ran parallel to it and to the north along the spur of the Gog Magog Hills (Fig. 28).

How far the integrated road system of Roman days survived through the long interval between the withdrawal of the Roman legions and the Norman Conquest, during which there was no central administration and probably little economic unity, it is very difficult to say. The survivals discussed would suggest that though undoubtedly stretches that served no local traffic decayed, some parts of the paved roads met the needs of the ordinary citizen, were kept in repair and remained in use. These useful stretches were sufficiently numerous to preserve the road pattern imprinted by the Romans. A map presented to the Bodleian Library, Oxford, by Richard Gough probably dates from the early fourteenth century, and the Roman roads certainly appear as the main lines of the system shown on it. William Harrison in his *Description of England* published in 1577 gives a list of the major thoroughfares of England and Scotland which suggests an integrated road system centred on London corresponding very largely with the Roman roads. John Ogilby's general map of the British roads again shows the same pattern in his day. The early sources may over-emphasise the nodality of London and the importance of the routes radiating from it. The Bodleian map or its original, from internal evidence, may have been drawn to illustrate the routes used by Edward I on his military journeys. William Harrison was an Essex man and may have known more about the routes leading out from London than of other roads, but John Ogilby's survey has some claim to completeness. It would seem therefore that London became the capital of the Norman kings in time to ensure that the state officials making long journeys on the king's business kept in being the through routes built to the frontiers of the province from Roman London, until such time as the commerce and trade of the whole country centred too on the administrative capital.

Medieval and Early Modern Communications

The Condition of the Roads

It is very difficult to gain a clear picture of the condition of medieval roads. Travellers say quite a lot about it, but the view of an individual about the goodness or badness of the roads depends very much upon the district in which he travelled, the season at which he travelled, and the way in which he travelled. The temperament of the traveller has also to be taken into account; some make light of even serious mishaps, others exaggerate every trivial discomfort and delay. On the whole travellers are apt to make the most of their woes; descriptions of accidents, diatribes against the shortcomings of man and of Nature make far more lively stories than records of success and paeans of praise.

Apart from the stretches of Roman road that remained in use, few roads were "built". The existence of the King's highway implied the existence of a right of passage, not of a road surface. People passed from place to place as and where they best could; the exact track varied. It might be lower down the slope on dry days and in summer to avoid unnecessary climbs or to be near springs for watering, but it might be forced higher up on wet days and in winter. If the usual track was rutted and holed the travellers skirted it on one side or the other. Where fields were unfenced, and especially if animals were frequently driven along that way, the road might be a very broad one. Tusser in fact in his *Five Hundred Points of Good Husbandry*, published in 1573, considers the encroachment of the public highways on the arable fields one of the great disadvantages of open land.

> "In Norfolk behold the dispaire
> of tillage too much to be borne:
> By drovers from faire to faire
> and others destroying the corne",

and again:

> "What footpaths are made, and how brode
> annoiance too much to be borne:
> With horse and with cattle what rode
> is made through every mans corne
> Where champion ruleth the roste
> there dailie disorder is moste".

William Harrison on the other hand takes the point of view of the traveller. He writes in 1587:

"Of the daily encroachment of the covetous upon the highways I speak not. But this I know by experience that whereas some streets within these five and twenty years have been in most places fifty foot broad according to the law, whereby the traveller might either escape the thief or shift the mire, or pass by the loaden cart without danger to himself and his horse; now they are brought unto twelve, or twenty, or six and twenty at most, which is another cause also whereby the ways be the worse, and many an honest man encumbered in his journey".

It is an interesting comment on this old controversy to read in *The Times*, 25 August 1949, that the Lancashire County Council propose that the standard width of their new main roads should be reduced from 120 ft. to 74 ft. "to help farmers, who were complaining that too much agricultural land was being taken".

The roads varied very much from region to region. This was in part because their upkeep was a local responsibility. Harrison describes the method of repair as it was in theory and in practice in his day:

"Wherefore by authority of Parliament an order is taken for their yearly amendment, whereby all sorts of the common people do employ their travail for six days in the summer time upon the same. And albeit that the intent of the statute is very profitable for the reparations of the decayed places, yet the rich to do so cancel their portions, and the poor so loiter in their labours, that of all the six, scarcely two good days work are well performed and accomplished in a parish on these so necessary affairs. . . . Sometimes also, and that very often, these days work are not employed upon those ways that lead from market to market, but each surveyor amendeth such by-plots and lanes as seem best for his own commodity and more easy passage into his fields and pastures".

The roads also varied very much on different soils. The unmade surface was quite satisfactory on well-drained but soft land, but much less so on stony ground or in waterlogged areas. The seventeenth- and eighteenth-century traveller was often well equipped with his guide books, itineraries and timetables. But many a pot boy must have laughed as he watched the stranger planning out the route, since he knew full well that the road marked so clearly in the *Britannia Depicta* disappeared into an impassable morass once across the parish boundary or off the chalky soil.

All are agreed that in the heavy clay lands the road surfaces were poor. William Harrison, *Description of England*, 1587, 2nd edition writes, "Now to speak generally of our common highways through the English part of the isle (for the rest I can say nothing), you shall understand that in the clay or cledgy soil they are often very deep and troublesome in the winter half". Reyce in his *Breviary of Suffolk*, the preface of which is dated 1618, thinks with delight of what will happen " by that time the invaders meet with our myrie soyle, our narrow and foul lanes ", and, fearing invasion, counts these bad roads as among the " commodities " or advantages of Suffolk's situation. The Wealden ways were notoriously bad : Daniel Defoe claims that " there I had a sight which indeed I never saw in any other part of England : Namely that going to Church at a Country Village not far from Lewis I saw an Antient Lady, *and a Lady of very good Quality I assure you*, drawn to Church in her Coach with Six Oxen, nor was it done but out of mere Necessity, the way being so stiff and deep that no Horses could go in it ". Arthur Young adds to his letters describing his *Six Months' Tour through the North of England* in 1770 one commenting on the state of the roads, and he has much to say about the dreadful condition of the roads across the wet soils of the Cheshire and Lancashire plain. He tells those

> " Who travel on any business but absolute necessity to avoid any journey further north than Newcastle (Newcastle under Lyme). All between that place and Preston is a country one would suppose devoid of all those improvements and embellishments which the riches and spirit of modern times have occasioned in other parts ".

From Warrington to Wigan

> "is a paved road, and most infamously bad. Any person would imagine that the boobies of the country had made it with a view to immediate destruction; for the breadth is only sufficient for one carriage; consequently it is cut at once into rutts; and you will easily conceive what a break-down dislocating road rutts cut through a pavement must be".

But north of Wigan to Preston the going he considers is even worse:

> "I know not in the whole range of language, terms sufficiently expressive to describe this infernal road. To look over a map, and perceive that it is a principal one, not only to some towns but even to whole counties, one would naturally conclude it to be at least decent; but let me most seriously caution all travellers, who accidentally purpose to travel this terrible country, to avoid it as they would the devil; for a thousand to one but they break their necks or their limbs by overthrows or breakings down. They will here meet with rutts which I actually measured four feet deep, and floating with mud only from a wet summer; what therefore must it be after a winter?"

Celia Fiennes tells in a rush of words what it may be like in winter, though she, travelling on horseback in 1698, complains much less bitterly of the road:

> "Preston is reckoned but 12 miles from Wiggon but they exceed in length by farre those I thought long the day before from Liverpoole; it's true to avoid the many mers and marshy places it was a great compass I took and passed down and up very steep hills and this way was a good gravel way; but passing by many very large arches that were only single ones but as large as two great gateways and the water I went through that ran under them was so shallow notwithstanding these were extreme high arches, I enquired the meaning and was informed on great raines those brookes would be swelled

to so great a height that unless those arches were so high no passing while it were so; they are but narrow bridges for foot or horse and at such floods they are forced in many places to boat it till they come to those arches on the great bridges which are across their great rivers: this happens sometimes on sudden great showers for a day or two in summer, but the winter is often or mostly so that there is deep waters so as not easily cross'd".

There may, therefore, be some justification for Arthur Young's warning: "I would advise all travellers to consider this country as sea and as soon think of driving into the ocean as venturing into such detestable roads". He does not wholly blame the soil: "only bad management can occasion such very miserable roads in a country so abounding with towns, trade and manufactures". The amount of traffic the roads had to bear, as well as the mossy soil posed indeed a difficult problem to the inhabitants.

The roads of the hilly districts though drier were by no means good: travellers tell of the difficulties they meet because of the narrow lanes, the steepness of the roads and the hard stoniness of the surface. Celia Fiennes journeying in Dorset and Devon says

"the ways very narrow, so as in some places a Coach and Waggons cannot pass; they are forced to carry their Corn and Carriages[1] on horses backs with frames of wood like pannyers on either side of the horse, so load it high and tie it with cords; this they do altogether the farther westward they go for the wayes grows narrower and narrower on to the Lands End".

In the Lake District Celia notes the same narrowness of the lanes:

"So went from Kendal to Bowness, six miles thro' narrow lanes: . . . here can be no carriages but very narrow ones

[1] In a later passage this is explained "all over Cornwall and Devonshire they have their Carriages on horses backs"

like little wheel-barrows that with a horse they convey their fewell (fuel) and all things else; they also use horses on which they have sort of pannyers some close some open that they strew full of hay, turff and lime and dung and everything they would use, and the reason is plaine from the narrowness of the lanes".

Arthur Young has worse to say of the narrow roads of Cleveland:

"From Newton to Stokesley in Cleveland . . . you are obliged to cross the Moors they call Black Hambledon over which the road runs in narrow hollows that admit a south country chaise with such difficulty, that I reckon this part of the journey made at the hazard of my neck. The going down into Cleveland is beyond all description terrible; for you go through such steep, rough, narrow, rocky precipices, that I would sincerely advise any friend to go a hundred miles about to escape it".

Arthur Young is hardly encouraging to his fellow travellers: the road to Askrigg "runs over the mountains, and is fit only for a goat to travel".

But even Arthur Young, demanding as he did a good smooth surface for his wide southern coach, did not consider all the roads poor, even in the North. In the Lake District the main roads please him. The road from Kendal to Windermere he records as a "Turnpike now making. What is finished, is as good, firm, level, a road as any in the world. I nowhere remember a better". From Kendal to Shap was "exceedingly hilly, and some very steep, but the road itself was excellent". From Shap to Penrith and on to Carlisle the roads are good, but about Carlisle he, like Celia Fiennes a century earlier, remarks on the little carts. The road from Carlisle to Brampton is "vilely cut up by innumerable little paltry one horse carts". The little carts that amused Celia Fiennes merely annoy Arthur Young: they have no business to make the going difficult for his stately coach. The road across Stainmore "is a most excellent one; firm, dry, level, and free from loose stones"; and for the most part the Great North Road from Doncaster to Stamford meets with approval.

On the downlands of the south travel was easier: Celia Fiennes talks of the road from Alton and Alresford to Winchester as " a good chaulky way ", and Arthur Young in his *Six Weeks' Tour through the Southern Counties of England*, published 1768, says that the road

> " from Salisbury to four miles the other side of Romsey, towards Winchester, is without exception the finest I ever saw . . . To management the goodness of the road must be owing; for fine as their materials are, yet I have in other roads met with as fine; but never with any that were so firmly united, and kept so totally free from loose stones, rutts and water: . . . To conclude the whole, it is every where broad enough for three carriages to pass each other; and lying in straight lines, with an even edge of grass the whole way, it has more of the appearance of an elegant gravel walk, than of a high road ".

The metropolitan roads by the end of the eighteenth century were also good: according to Arthur Young, " Next to this uncommon road, (the Salisbury-Romsey road), the great north one to Barnet, I think, must be ranked. Then the Kentish one: and the others to Chelmsford and Uxbridge succeed ".

The Traffic on the Roads

In the Middle Ages people moved about a good deal. The king toured his realm with his knights and courtiers. The royal travellers could never stay long in one place; they literally ate themselves out of house and home very quickly. The lords with their households " perambulated " their manors, it was easier sometimes to move people than food, so they ate their way around their lands. Soldiers were often on the move, and soldiers then as now meant able-bodied civilians called upon " to muster " for service at a given place. The subjugation of Wales, the fighting in Ireland and Scotland and in France meant the movement of men and of supplies to the battlefields or the ports of embarkation. Less organised but at times hardly less numerous were the pilgrims who flocked in search of health for body or soul to the shrines of, for example, our Lady at Walsingham, St. Thomas at Canterbury, St. Joseph at Glastonbury or St. Winifred at Holywell. Others travelled to the ports *en route* for Santiago de Compostela, Rome or even Jerusalem. The number of village churches that contain Crusaders' tombs reminds us of the numbers of folk who made those part religious, part military and often no doubt, too, part trading journeys to the Holy Land. It is true that for the satisfaction of everyday wants people did not need to travel very far in lowland England; in the days before the railways the nearest market town was but four or five miles away and there, weekly, corn and wool, butter and cheese would be sold by the country-folk who bought with the proceeds scythes and ploughshares, cloth and shoe-leather, pots and pans, crockery, salt and perhaps the commoner spices. But there was also trade and commerce on a much wider scale. Wool merchants, English and foreign, travelled around the countryside buying the wool clip from manor and farm, and the good wool was transported to the big wool fairs or to the ports licensed to export it. Cloth merchants moved about distributing wool to spinners, yarn to weavers, cloth to fullers and dyers and to markets, fairs and ports. All and sundry flocked to the fairs. William Harrison says, "there is almost no town in England but hath one or more such marts held yearly in the same although some of them I must needs confess (have) little else bought or sold in them more than good drink, pies and some pedlery trash". He then gives "a calendar of the greatest, sith I cannot, or at least wise care not, to come by the names of the lesser". His table contains some 360 fairs: indeed plenty. Many of these so-called 'great' fairs were only of local importance; but in the twelfth and thirteenth century the four fairs of Northampton, St. Ives, Boston and Winchester attracted merchants not only from all parts of the British Isles but from Europe too. In the second half of the thirteenth century even their heyday passed; commerce was becoming more and more concentrated on a few centres, the staple ports of Hull, Lynn, Bristol and above all London, which would account for the fact that, as soon as we have anything even faintly approaching a reliable map showing communications, all roads lead to London. One fair of not more than the second rank in the thirteenth century is by the fifteenth by far the most, in fact the only, important English fair—Stourbridge outside Cambridge. The date and the circumstances of the foundation of Stourbridge Fair are unknown, for all that Thomas Fuller in his *History of Cambridge* (1655) gives a pretty fiction.

> "A clothier of Kendale", he tells, "casually wetting his Cloath in that water on his passage to London exposed it there to sale, on cheap terms as the worse for the wetting, and yet it seems saved by the bargain. Next year he returned again with some other of his Townsmen, profering drier and dearer cloath to be sold: so that within a few years hither came a Confluence of Buyers, Sellers and Lookers-on, which are the three Principles of a Fair."

It was already probably a flourishing fair in the twelfth century and in 1211 its tolls were given to the Leper Hospital nearby; it remained a flourishing fair until the eighteenth century, when Defoe tells us that "it is not only the greatest in the whole Nation, but I think in Europe; nor are the Fair at Leipzig in Saxony, the Mart at Frankfurt on the Main, or the Fairs at Nuremburg, or Augsburg, reputed in any way comparable to this at Stourbridge". There is a footnote however in the Fifth Edition of Defoe's *Tour* (1753) which runs, "this Fair is pretty much dwindled since this Account of it; tho' it is still very considerable". No better idea can be gained of the amount of merchandise that could be collected by road and river at a given place, and a country place at that, than by reading Defoe's description of Stourbridge Fair. It is too long to quote, but it is to be found in Volume One of the *Tour through the Whole Island of Great Britain*, in any of the first five editions. Fuller emphasises the importance of both the land and water routes: "It is at this day", he writes, "the most plentifull of Wares in All England, (most Fairs in other places being but Markets in comparison thereof:) being an Amphibion, as well going on Ground as Swimming by Water, by the benefit of a navigable river".

All the travellers of medieval days who could afford it rode on horseback, they could negotiate the narrow bridges, ford safely all but the swifter and deeper streams, and pick their way along the best parts of the tracks. The ordinary merchant not in a hurry would cover 20 to 30 miles a day, but much greater distances could be made when there was urgency. Early in the sixteenth century post stages were established; at each stage, usually a chosen inn, relays of horses were kept ready to ride on to the next post. On official business horses could be hired for 2½d. a mile, on private business for 3d. a mile, and a post boy, also mounted, to ride with the traveller and bring back the horse cost 6d. To ride "post-haste" was to ride about 10 miles an hour, and a journey of 70 to 150 miles could be made in one day if necessity was a sufficient spur and the road good.

Pilgrims, soldiers, pedlars, and all other poor men most often walked. So did the drovers of animals. The only easy way, the only possible way to take meat to market over any distance was "on the hoof"; it was not only cattle, sheep and pigs that moved literally on the hoof, but geese and turkeys, ducks and even hens were to be seen footing it into London and other big towns.

All traffic could not walk or ride, though a surprising variety and bulk of goods did ride in panniers or in bales on the backs of pack-horses. Wheeled vehicles were, however, also used. The Britons possessed wheeled carts before the Romans came. Much farm work involved carting; dung, hay, corn were all bulky products. Some heavy goods had to be moved over long distances; coal, ore, building stone, timber, and the like occurring locally had a wide market. In the lists of tolls granted to towns to help to pay for the building of walls, the paving of streets or the repair of bridges cart-loads of this and that are frequently mentioned. Norwich in 1297 was allowed to charge ½d. for "every cartload of firewood or timber coming weekly" and 4d. "of every cartload of sea-fish". Yarmouth in 1346 charged 1d. on "every cartload of turves coming or going weekly" and Colchester in 1399 took 1d. in duty payable at the quay for every cartload of fullers earth.

The amount of wheeled traffic steadily increased; before the end of the sixteenth century coaches and waggons, both private ones and those plying for hire, were commonly to be seen not only about the streets of London and other big towns but trundling along the country roads. Dr. John Dee, the cosmographer and geographer appointed Warden of Manchester College in 1595, sent his wife and family from Mortlake as far as Coventry in a coach and employed " Percival the Lancashire carrier " to move his furniture. In the later decades of the sixteenth century " Thomas Hobson, the carrier of Cambridge, by the help of common prudence, and a constant attention to a few frugal maxims, raised a much greater fortune than a thousand men of genius and learning, educated in that University, ever acquired or were even capable of acquiring ". He plied regularly between Cambridge and London, carrying most goods by packhorses, but also conveying goods, women, children and sometimes scholars too in waggons.

Coaches, four-wheeled vehicles for passenger transport, were introduced from France or Holland to England sometime early in the sixteenth century, and their popularity increased rapidly especially in London. In 1601 a Bill was brought into the House " to restrain the excessive use of coaches " in England and one Fynes Moryson describing his journeys in England and on the Continent in his *Itinerary* published 1617 says " the streets of London are almost stopped up with them". Traffic jams do not belong only to modern days. In 1637 there were two weekly coaches running between St. Albans and London, the first regular service of its kind yet recorded. By 1658, judging by advertisements of that date, there was regular stage-coach communication from London as far afield as Devon and Cornwall in the south-west, Newcastle and Edinburgh in the north, and Chester, Wigan and Preston in the north-west. The ordinary rate of travel by stage-coach seems to have been between 20 and 30 miles a day, though on good roads in summer months perhaps 60 miles or more might be covered. Edward Chamberlayne in his *Angliae Notitia*, or the *Present State of England* (1669) tells of the new way of travel :

> " there is of late such an admirable commodiousness, both for men and women, to travel from London to the principal

> towns of the country that the like has not been known in the world, and that is by stage coaches, wherein anyone may be transported to any place sheltered from foul weather and foul ways, free from endamaging one's health and one's body by hard jogging or over violent motion on horseback ".

If Arthur Young tells the truth about his journeyings a century later the comfort of the new way did not prove to be all it was cracked up to be; and many continued to journey more rapidly if more strenuously by post-horses.

Improvement of the Roads

Stage-waggons and stage-coaches did not replace but were added to the trains of pack-horses and the strings of post-horses using the roads. At first wheeled traffic probably brought about an improvement in road surfaces, but the increasing volume and weight of the traffic may have caused an absolute deterioration, as it certainly caused louder and louder complaint; and more attention had to be given to the state of the roads. Early efforts went the wrong way to work and attempted to limit the traffic rather than improve the roads. Acts were passed limiting the breadth of cart wheels, the size of carts, the load of the carts, and the number of draught animals that might be used to draw the carts. With the increased pressure on transport facilities resulting from growing industries and increasing trade and commerce, most of these Acts were dead letter Acts from the moment they appear on the Statute Book. A more realistic approach was an attempt to make those who caused undue wear and tear upon the roads put in extra work upon them. The iron-masters of the Weald for example had special obligations. It came to be realised that the two things most needed were money for repairs and some standardisation from area to area of at least the main roads. In 1654 an Act permitted the parishes to raise the rate payable for the upkeep of highways; in 1656 a Surveyor-General for the whole country was appointed; and in 1663 an Act providing for the repair of the Great North Road by collecting tolls from travellers was the beginning of the Turnpike system. The Turnpike system which accepted the principle that the users of the roads should contribute to their upkeep did much not only to improve the roads but to make the main roads more even in quality throughout their length. A poor parish was no longer confronted with the impossible task of maintaining the surface of a busy highway that happened to pass through it. The turnpiking of roads was a slow and piecemeal business; it was costly and there was much opposition to tolls. More than twenty Acts were passed and over a century elapsed before the arterial road London to Carlisle was turnpiked throughout its entire length.

The Turnpike system marked a great improvement in the administration of road upkeep and thus in the effectiveness of the repairs carried out, but it did not in itself mark any advance in the technique of road construction. Roads were still made even at the end of the eighteenth century without any proper foundations; road material was merely laid upon the natural surface and heaped higher in the middle to help drainage. Wheeled traffic soon wore down the camber and cut the surface into ruts. There was undoubtedly some justification for Arthur Young's complaint that "the only mending it in places receives, is the tumbling in some loose stones, which serve no other purpose but jolting a carriage in the most intolerable manner". But the day of technical improvements was in sight; John Metcalfe, the blind engineer of Knaresborough, constructed in the 1760's between Harrogate, Knaresborough and Boroughbridge the first scientifically laid road sections since Roman days. In the last two decades of the eighteenth and the first two of the nineteenth century two Scots, Thomas Telford and John Macadam, stirred by the obvious deficiencies of the northern road system, carried out experiments in construction there which were to revolutionise road making everywhere. Both insisted on good drainage and a smooth surface, but to secure these Telford advocated solid foundations, Macadam a waterproof covering. Macadam's method was the quicker and cheaper way of achieving the desired end and so our roads are "macadamised" today, and fulfil, after a partial eclipse in the nineteenth century, as important a function in the transport system of our land as they did in the Middle Ages.

Transport by Water : River Traffic

However, even in the Middle Ages the roads did not carry all the traffic ; the rivers and the coastal waters carried a great deal. Compared with travel by land, travel by water was slow. Professor Stenton in an article on "The Road System of Medieval England", in *The Economic History Review*, Vol. VII, 1936, re-tells from an Exchequer record an interesting travel story. In 1319 King Edward II invited the scholars of King's Hall (later Trinity College) Cambridge, to spend Christmas with him at York. The elder ones set out on hired horses and rode the 151 miles to York in five days. The younger ones on 20th December were packed off down river in boats. In two days they reached Spalding ; the third day they rode from Spalding to Boston, and there again they embarked in a big boat and set off up the Witham. It must have been, one imagines, a rather dispirited group of little boys who disembarked on Christmas Eve at Lincoln after four winter days spent on Fenland rivers and still far from the expected splendours of Christmas festivities at the Royal Court. However, on Boxing Day off they set in two boats down the Foss Dyke, a canal probably of Roman date recut in 1121, from Lincoln to Torksey ; they transferred the next day to a bigger boat and on 28th December after eight days' travelling arrived in York.

The rivers were perhaps not often the chosen routes for passenger traffic but for heavy and fragile goods (the youth of the King's Hall scholars may have placed them in this latter category), the rivers were used wherever possible. When coaches and waggons were introduced, the watermen feared that the rivers would lose their trade. John Taylor, the water-poet, cried out :

"Carroaches, coaches, jades and Flanders mares,
Do rob us of our shares, our wares, our fares :
Against the ground we stand and knock our heels,
Whilst all our profit runs away on wheels :"

But he need not have worried ; it was not on the wheels of his day that the profits of the boatmen were to run away.

Large timber could only be cut profitably within a few miles of navigable water ; coal had a market only within a short radius of the mines and the waterways ; the builders of cathedrals and castles, of monasteries and manors as well as of farms and cottages had to make do with local material unless they could transport better by water. As wood became scarcer and thus charcoal dearer, as population increased, as industries developed, coal came more and more to be the most important commodity carried on coastal and inland waters. But as the coal so carried has disappeared in smoke long ago it cannot now be used as an index of the communication system. Building materials however have a more permanent physical existence. It is an interesting exercise to examine carefully the materials of which a church is built with the problem of transport in mind. If timber or brick, was the area known to be well wooded ? If not, where was wood to build, or to fire the brick brought from ? Or was it coal that fired the kilns ? If stone, is it local ; if not, where does it come from ? A great deal can thus be learnt of the possibilities of transport by water. The limestones of the Midlands were excellent building stone ; they are found far from their outcrop in buildings all around the Fenland, in Lincolnshire, Cambridgeshire, in the west of Norfolk and Suffolk. Big and wealthy villages sometimes built the whole church of limestone ; the less wealthy used flints from the local chalk and boulders from the local glacial drifts for walls and the stone for towers and corner stones. The numerous round towers of mid-Norfolk and Suffolk reflect not only the absence of good local building stone, but also the distance of those churches from navigable water.

River traffic was not only a traffic in heavy raw materials. Cambridge had special hithes for salt, flax, and corn as well as for coal, and up-river came too every fenland product including, one of the most important among them, " 1500-2000 firkins of Butter every week, from Norfolk and the Isle of Ely which is sent by waggons to London ". The Yorkshire Ouse and its tributaries carried in addition to coal much wool and cloth, corn and butter and bacon, wine and a great quantity of imported groceries. Salt was of especial importance on the Weaver and Mersey ; and the Severn, in addition to the great traffic in coal and iron, carried down-river salt, corn, cheese and fruits, and up-river wine, sugar, oil, tobacco, groceries and dyestuffs. The trade of the Thames was very great, foodstuffs and cloths down to London, and upstream samples of almost every product of the then known world made in or imported to London.

England possessed in her major rivers and her numerous sheltered harbours around the coast an excellent system of natural waterways. Defoe estimates that in his day about 1,160 miles of the rivers were navigable; by then in general only the central areas of Wales, the Lake District and the Pennines were more than fifteen miles or one day's carriage by land from navigable water (Fig. 51). In medieval times, when sea-going vessels were of shallow draught, towns far inland like York, Doncaster, Cambridge, Norwich and Gloucester ranked as sea ports; and river vessels plied on many yet smaller streams. Changes in time worked both to decrease and increase the length of navigable waterways. The increasing size of boats, silting natural, and artificial by tipping out ballast to lessen the draught which was especially apt to take place where water was already shallow, the construction of dams and waterwheels for power, and here and there drainage schemes put some stretches of river and some ports out of use. On the other hand throughout the seventeenth and eighteenth century constant work, reflected in numerous River Navigation Acts, was going on to deepen shallow places, remove dams and wheels from the main river course and even to cut short artificial channels to get round bad stretches.

Changes in the distribution of available natural resources, particularly timber, changes in the evaluation of raw materials, particularly the increasing importance of iron and coal, changes in commerce with expanding world horizons and particularly the growth of a colonial empire, changes in industry and agriculture, all press on changes in the arteries of trade and commerce. The steady growth of industry, the improvements in agriculture in turn depended upon increasing capacity to carry information, raw materials, and finished goods in increasing bulk with increasing speed at decreasing cost. Many clothiers leave money in their wills for road improvements, testifying to their

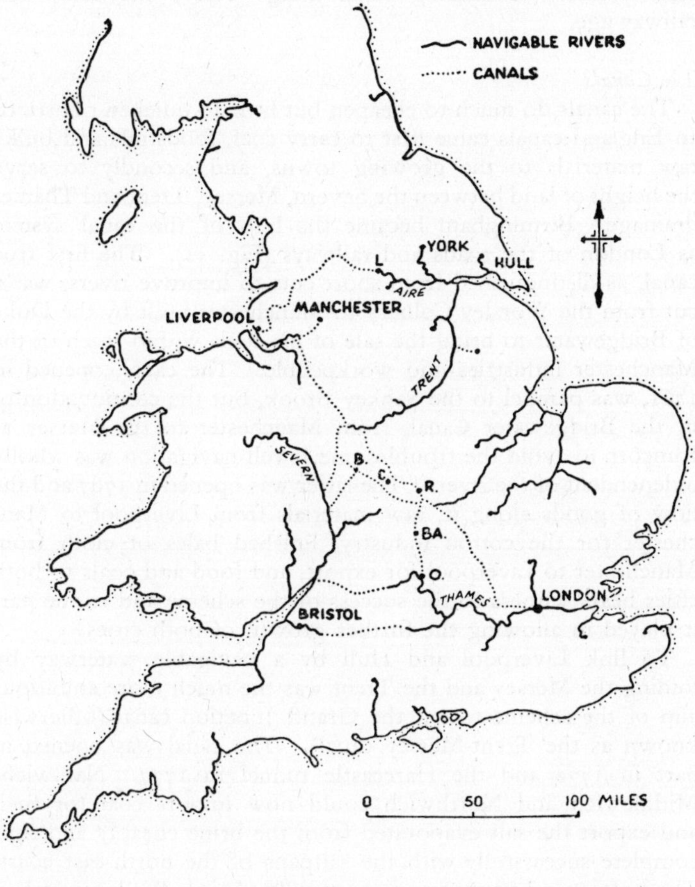

Fig. 51 The Principal Inland Waterways of the late Eighteen Century England L=Leeds ; B=Birmingham ; C=Coventry ; R=Rugby ; Ba=Banbury ; O=Oxford.

grasp of this need; it was partly difficulties of transport that killed the Wealden iron industry and limited the Midland iron manufacturers to making small things before the canal and railway age.

The Canals

The canals do much to cheapen but little to quicken transport. In England canals came first to carry coal, foodstuffs and bulky raw materials to the growing towns, and secondly to serve the height of land between the Severn, Mersey, Trent and Thames drainage. Birmingham became the hub of the canal system as London of the roads and railways (Fig. 51). The first true canal, as distinguished from short cuts to improve rivers, was a cut from the Worsley Colliery to Manchester built by the Duke of Bridgewater to bring the sale of his coals within reach of the Manchester industries and workpeople. The canal, opened in 1761, was parallel to the Sankey Brook, but the continuation of it, the Bridgewater Canal, from Manchester to the Mersey at Runcorn to avoid the troublesome Irwell navigation was wholly independent of the river. The latter was opened in 1767 and the flow of goods along it, raw materials from Liverpool to Manchester for the cotton industry, finished bales of cloth from Manchester to Liverpool for export, and food and coals to both cities bears witness to the success of the scheme and to the part it played in allowing the further growth of both cities.

To link Liverpool and Hull by a navigable waterway by joining the Mersey and the Trent was the much more ambitious aim of the scheme to cut the Grand Junction canal (otherwise known as the Trent-Mersey canal). The canal was opened in part in 1772 and the Harecastle tunnel in 1777. Nantwich, Middlewich and Northwich could now import coal for fuel, and export the salt evaporated from the brine cheaply and thus complete successfully with the saltpans of the north-east coast. The pottery industry grew very rapidly, Josiah Wedgwood had a great interest in the canal scheme. In 1760 it was said that the pottery industry employed about 7,000 people " almost as rough as their roads " ; in 1785 (eight years after the canal was opened) Wedgwood giving evidence in Parliament said 15,000 people were employed. In 1781 John Wesley in a letter from Stoke-on-Trent writes, "How is the whole face of this country changed in about twenty years. Since which inhabitants have continually flowed in from every side. Hence the wilderness is literally become a fruitful field. Houses, villages and towns have sprung up; and the country is not more improved than the people". The change was not all due to the canal, but remembering Arthur Young's strictures on the roads north of Newcastle-under-Lyme in the 'seventies, water transport must indeed have proved a boon. It allowed cheap carriage of bulky china clay from Devon and Cornwall, flints from Hull and the east coast counties, coal from local pits, thus cutting out the high cost of carriage in panniers on horseback from the head of river navigation of the Weaver or the Severn. It allowed foodstuffs to be supplied by a wide area, and destroyed the monopoly of the local grain merchants. It allowed the safe carriage of the fragile finished products, so unsuited to bumping along the deeply rutted roads.

Not only were the inland waterways of the Midlands linked to Liverpool and Hull but also to Bristol by the opening in 1772 of the Staffordshire-Worcester (or Wolverhampton) Canal from Haywood on the Trent to Stourport on the Severn (Bewdley having refused the honour of the terminus and with it the opportunity for rapid "development"). In 1782 Thomas Pennant in his *Journey from Chester to London* writes:

> "Notwithstanding the clamours that have been raised against the undertaking [The Grand Trunk Canal] in places through which it was intended to pass when it was first projected, we have the pleasure now to see content reign universally on its banks, and plenty attend its progress. The cottage, instead of being half covered with a miserable thatch, is now secured with a substantial covering of tiles or slates brought from the distant hills of Wales or Cumberland. The fields, which before were barren, are now drained, and, by the assistance of manure conveyed on the canal toll-free, are clothed with a beautiful verdure. Places which rarely knew the use of coal are plentifully supplied with that essential article

upon reasonable terms ; and what is of still greater public utility the monopolisers of corn are prevented from exercising their infamous trade ; for the communication being opened between Liverpool, Bristol and Hull, and the line of the canal being through countries abundant in grain, it affordeth a conveyance of corn unknown to past ages ".

To link this system to London was to break finally the isolation of the Western Midlands. Acts were passed in 1768 and 1769 approving the construction of a canal from Coventry to Fradley Heath on the Grand Junction Canal and a canal from Coventry via Rugby and Banbury to the Thames at Oxford, but opposition from varied vested interests perhaps helped to increase costs, for certainly money troubles delayed the work and the through waterway was not complete till 1790. Yarmouth merchants voiced opposition to this scheme on the grounds that the wide distribution of Midland coal over the area in general and to London in particular would ruin the Newcastle coasting trade in coal ; Yarmouth was interested as the port supplying East Norfolk by the Yare and Bure navigation ; she was in fact with King's Lynn the leading importer of Newcastle coal after London. In 1730-31 London imported 217,000 chaldrons of coal from Newcastle, King's Lynn 13,403, Yarmouth 12,657, followed by Hull 3,264.

In the north canals were projected, sanctioned and cut across the Pennine divide to link the Yorkshire and Lancashire manufacturing districts, and to provide an outlet to the growing port of Liverpool with its African and American trade for the woollen manufactures of the West Riding. The Leeds-Liverpool canal via Wigan, Blackburn and Skipton through the Aire gap, was opened in part in 1777, throughout in 1790 ; the through route was not wholly successful ; there was too little water in the dry season to work efficiently so many locks and sometimes two or three days would be spent in some locks. Dr. Aiken, the topographer of the Manchester district, commenting on the fact that no less than three canals crossed the Pennines, hopes, "long may the districts of Lancashire and Yorkshire remain the centre of a trade capable of maintaining these mighty works". He did not foresee that the day was not far distant when "these mighty works" would decay, not because trade failed but because it grew beyond their capacity. It has been said that the canals were "too soon, too many and too narrow". There were too many; the country suffered a "canal mania" in the 1790's, when, encouraged by the success of a few canals where they served to release potentialities pent up for lack of means of transport, many were built from nowhere in particular to nowhere in particular and had from the first no hope of attracting trade. Some of these canals derelict from the beginning afford excellent lessons in the study of location; they emphasise the fact that a single opportunity or amenity is not sufficient to determine the growth of trade and industry. Secondly, many *were* too narrow and perhaps, in that, they were too soon. They were too narrow to allow for increase of traffic, to allow laden boats to pass easily, and to allow for wash when boats were moving quickly. It was found that movement between four and eight miles an hour created a wash which damaged the banks; movement 10 to 11 miles an hour was again safe. This latter speed needed steam power and the narrow canals and especially the small size of the locks hampered the use of boats big enough to run economically with engines. If the canals had been cut when the possibilities of steam engines were more fully realised, they might possibly have been made bigger, but to wait might have been to miss the moment. Britain gained much by being first in the industrial field, and improved water transport helped to gain that lead. Transport by canal was undoubtedly slow; it took seven to ten days from London to Bristol by the Thames-Kennet-Avon route in 1835, London to Liverpool was seven or eight days and nights in 1831, and even Birmingham to London using fly boats in 1832 took four days. But the cost of transport was much reduced; in 1792 to move goods from London to Reading by road cost 33s. 4d. a ton but only 10s. a ton by water; from Manchester to Birmingham £4 0s. 0d. a ton by road and £1 10s. 0d by canal, and in 1813 it was said that nowhere did canal carriage cost more than half land carriage. Not only many old factories were well served by the canals but many new ones were made:

a canal side site was a favourite choice for new buildings, land was cheaper than in a town and transport by canal allowed town products as well as raw materials to be brought to the site. By the time the canals lost their monopoly many of these canal side sites had grown so important that it was worth while building railways to feed them; others languished for a time to revive when road transport grew to meet their needs; yet others decayed. In discussing the location of industry or in examining the regional pattern of agricultural improvements, the conditions of the canal era may in certain areas provide an all-important clue.

The Railway Age

The coming of the railways is not the sudden event that it often seems; where roads were soft or bumpy and materials heavy the use of railways to ease the passage of carts regularly carrying heavy goods along a given route was early. Most of these early railways used wooden rails and most of them were built to carry coal or iron; they were commonly used in the Newcastle area to bring coal from the mines to the river in the later seventeenth century. Improvements were gradual: wooden rails wore easily, especially on curves, so iron plates were fixed to the rails like tackets to shoes to protect them; later cast-iron rails were used; these cracked, so wrought-iron rails were substituted; and eventually when steel became reliable enough in quality and cheap enough in quantity, steel rails replaced iron rails. Wheels changed too. Instead of wooden wheels, iron wheels were tried and iron wheels with an inner flange to keep them on the rails. Horses at first provided the power. The early railways were secondary feeding lines to the nearest navigable water. They were not considered as main lines of transport in themselves. It was the combination of "railways" with steam power that wrought the revolution. Enthusiasts prophesied great things, the experts as ever were cautious. "It is far from my wish to promulgate to the world that the ridiculous expectations or rather professions, of the enthusiastic specialist will be realised, and that we shall see them travelling at the rate of twelve, sixteen, eighteen or twenty miles an hour. Nothing could do more harm than . . . the promulgation of such nonsense ". (Nicholas Wood.) But on 27 September 1825 along the Stockton-Darlington line, there for all to see steamed an engine drawing a train of about eighty tons weight up an incline at from ten to fifteen miles an hour.

Fig. 52. The Railways of Southern Britain in 1850.

So rapidly did railways develop that by 1850 the general network in England was complete (Fig. 52). The railways like the turnpike system of roads, and the Roman roads before them, served as main lines of traffic and were thus early integrated into a system. Because of the moment of their construction, they centred upon London, the commercial and financial as well as the political capital. There were many lines built afterwards; too many; like some canals, some railways were constructed apparently without any thought being given to what was to be carried by them, and like the redundant canals these railways attracted no traffic and earned no dividends.

The railways not only cheapened but enormously increased the speed and range of communications and the volume of traffic that could be handled. The railway rates were soon cheaper than the canal rates; in 1836 the rates for raw sugar from London to Birmingham were by canal 40s. a ton, in 1842 by railway 37s. 6d. and in 1880 20s. a ton; in 1811 the freight on valuable goods from Birmingham to Manchester by canal was 55s. a ton, by land about 80s. a ton and in 1842 by rail 25s. a ton. The competition of the railways reduced the cost of canal transport and rapidly put the long distance waggoners and stage-coach proprietors out of business altogether, although local feeder services to the railway stations may have increased. Time was even more important in considering passenger transport, though railway travel for passengers was cheaper too: in 1844 it is said that the fare from Manchester to London by canal was 14s., by coach 30s. and by train (third class) 25s. but the saving in time was very great, the canal journey took 5 days, the coach journey about 20 hours and the railway journey 10 hours. Movement of people and goods became far freer; the migration of population was no longer in the 1860's and 70's, as it had been in the 30's and 40's, largely short distance movement from the country to the nearest market town, people now moved more freely to meet demand and most freely along lines of easy journeys by rail. The migration from all parts of the country into London, for example, was much quickened. Locally, the railways made a great difference; roughly every other market town decayed. The market towns had been spaced to provide a business and shopping centre within an easy day's journey by horse; once the railways linked the villages to the towns, then fewer and bigger markets grew where railways met; every reader must be able to furnish local examples of this change. When for short distance travel motor cars and buses replaced the railways the pattern of country markets did not change much, though here and there an otherwise convenient centre that by accident was not well served by the railway net has grown quickly in this "motor" age. With the era of competition between rail and road, modern conditions are present.

The present pattern of settlement of all the countries of Europe has in its essential form been established before the railway age, and can, like that of Britain, only be appreciated in the light of this fact. In America the change from the most primitive system of transport dependent upon human porterage to the most highly developed one dependent upon the steamship, the railway engine, the motor car and the aeroplane has occurred during the period of the first settlement. There is literature in plenty, and stirring literature too, describing conditions of travel by land, river, canal and railway in the new continent. The first great expansion westwards, the crossing of the Alleghenies, the trek of the settlers up the Hudson and the Mohawk to the Ohio plains, through the Cumberland Gap down the Kentucky and the Tennessee, was carried out on foot and in waggons. To read of this helps to make come alive the scene in Britain and in Europe as colonists there too first trod the yet untrodden ways. In contrast, the settlement of the West following the building of the railway lines presents a very different picture; areas of first colonisation are closely related here not to the natural routeways but to the lines of the railway concessions. In many other places the striking, indeed often spectacular, changes brought by new means of communication and transport are evident: railway transport brought great opportunity and grave problems to the Rand, ships with refrigerated holds transformed the prospects of pastoral farmers in the Argentine and New Zealand. In these new countries the effects of the rapid transformation within a generation of an area from a wilderness occasionally visited by man to a cornfield, a mine- or oil-field in easy reach of the most sophisticated and sensitive markets may be examined. Here, if anywhere, as a result of changes of transport, the geography of a country can be watched in the making.

Map Exercises

Suggestions of maps that the reader may make of his own area to illustrate this chapter.

1. Map of the roads and railways about 1850.
2. Map of the canals, navigable rivers and main roads about 1800.
3. Map of roads shown in Ogilby's *Britannia* of 1675 or in an early revision of it, e.g. John Senex' *The Travellers' Guide* (1699) or Emanuel Bowen's *Britannia Depicta* (1720).
4. Map showing
   (a) medieval bridges
   (b) place names containing elements strat, street, etc.; bridge; ford.
   (c) archaeological finds indicating Roman roads and prehistoric tracks.

CHAPTER IX

THE PLACE OF HISTORICAL GEOGRAPHY IN THE GEOGRAPHICAL SYNTHESIS

GEOGRAPHY has an essential unity of theme: in fact Vidal de la Blache taught that " what geography, in exchange for the help which it receives from other sciences can bring to the common treasury, is the art of not dividing what nature brings together ". It would seem fitting therefore in this concluding chapter to consider the part that Historical Geography takes in this synthesis that is a whole.

I

Every geographer, whatever the stage of his study, must at one and the same time be conscious of the immensity of the whole, yet must consider it part by part if he is to examine his material in sufficient detail to understand the complexities involved and see exactly how the parts from smallest to largest interact and interlock. Close observation and detailed study provide a firm basis of knowledge from which alone broad and illuminating generalisations can be framed. The geographer then must find a way to subdivide the study of his subject for convenience, although he cannot subdivide the subject itself. Geography, the study of place, can be logically subdivided into the distribution and location of the natural phenomena, subdivisions which have links with the Natural Sciences; and of the distribution and location of the phenomena of human societies, which have links with the Social Sciences. The geography of the natural world when subdivided into Physical Geography, concerned with those elements of place belonging to the physical world and having links with the Physical Sciences, and Biological Geography or Biogeography, concerned with

GEOGRAPHY : THE STUDY OF PLACE.

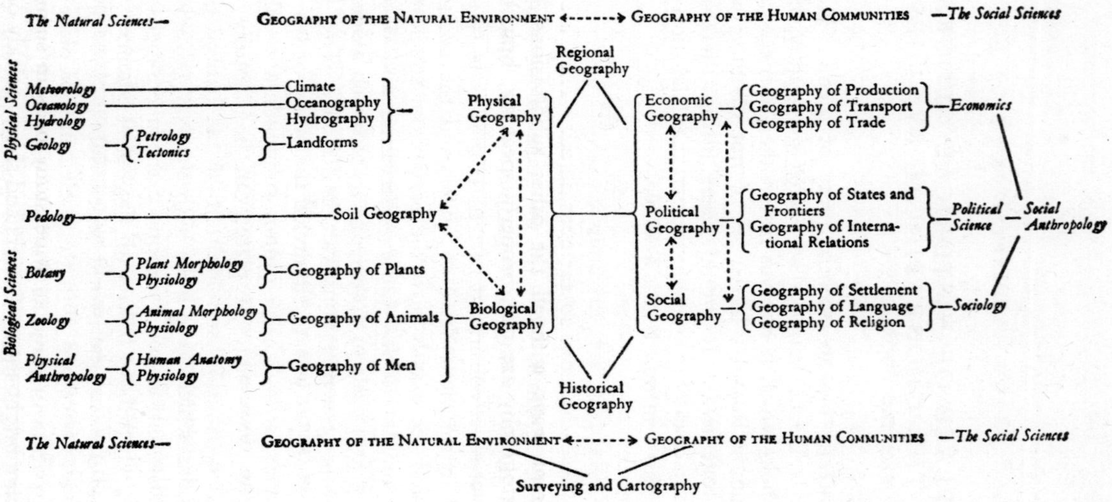

those elements of place belonging to the living world and having links with the Biological Sciences. Soil Geography stands between these two groups: the content of the soil is partly mineral and climate determines the major soil groups, but the organic content and the living soil organisms are an essential element, and soil development is closely linked to plants, animals and men. The geography of human societies can be divided parallel with the Social Sciences: into Economic Geography, concerned with those elements of place that result from the work man does to earn his living and closely linked with the study of Economics; into Political Geography, concerned with those elements of place that result from the work man does to govern his communities and closely linked with the study of Political Science; and Social Geography, concerned with those elements of place that man creates by living in communities and closely linked with the study of Sociology.

Surveying and Cartography, Regional Geography, and Historical Geography are not branches of geography of the same order as the divisions so far considered. They are concerned in different ways with all the other subdivisions. Surveying and Cartography are not, properly speaking, branches of geography at all; they are technical sciences, arts too. All geographers must be thoroughly versed and practised at least in their simple procedures, for the map is both the apparatus in constant use to make geographical investigations and the record of investigations made. Map making and map reading are with the geographer first, last and all the time, and are second in importance only to investigation and exploration on the ground itself. Regional Geography is the study of any or all the divisions with reference to a particular area. A region is a part of the earth's surface that has obvious unity but is not necessarily a political unit or even possessed of a place name. Lands lying about the Mediterranean Sea, for example, all have hot dry summers and warm wet winters; they have strong and varied relief, high mountains and deep valleys, hill and plain in quick succession; the rivers are full streams in winter, trickles or pools in wide stony beds in summer; many plants have bright showy flowers and shiny, prickly or thickened fleshy leaves; the people for the most part have small bones, dark hair and dark eyes, and long and narrow skulls; the crops include vines and olives and oranges; cultivation is a repeated shallow working of thin soil; industrial development is slight; also in this, the home of Greek and Roman civilisation, the nation states are yet young. There is an obvious likeness between southern Spain, southern Italy and Sicily, Greece, northern Tunis, Algeria and Morocco, and the eastern fringes of Asia Minor, and so they are grouped together as "the Mediterranean Region". A region is sometimes studied as a "natural" unit. The natural region is as it is as the result of the interaction of all those elements of place, including man himself, that are not man-made, however much altered by him. In contrast to but including this, there is the "geographical" region; this means the region as it is as a result of the interaction of *all* the elements of place including those that are man-made. The study of the Mediterranean as a natural region includes the study of its climate, relief, the wild plants and animals, and the people considered as biological groups. The study of the Mediterranean as a geographical region includes also the study of its villages and towns, its agriculture, its industries, and its political organisation, too, in so far as this affects the place.

Historical Geography is the study of any or all the subdivisions with reference to a particular period in the historic past.

II

If every historical geographer must be versed in all parts of geography, every geographer must be to some extent a historical geographer. No biographer, to return to the original analogy, would consider his man in his maturity and old age without at least a backward glance at his youth and childhood.

The physical geographer, concerned as he is with the physical scene that, however rapidly it is changing in geological time, is in general relatively stable in historical time, does find that some of its features show important changes in evolution and development within the historic period. Some coasts, for example, where rocks are not resistant to erosion and structures are weak, change rapidly. The changes that have taken place on Scolt Head Island since the eighteenth century are discussed in *Scolt Head Island*, edited by J. A. Steers. Contemporary maps provided much of the information on which the study of the geographical evolution of the island is based. The study of the development of Dungeness (W. V. Lewis. "The Formation of Dungeness Foreland", *Geographical Journal*, Vol. LX, 1932 and W. V. Lewis and W. G. V. Balchin, "Past Sea Levels at Dungeness" *Geographical Journal*, Vol. XCVI, 1940) is another good example of the way in which the historical approach may help in solving a problem in physical geography. Rivers may change their courses swiftly and suddenly, volcanoes may appear and disappear almost overnight, and sand dune belts may grow and decay rapidly; rewarding studies in historical geography might be made of these features where records of the changes can be found.

There is much speculation about climatic changes in historic time and illuminating studies await the climatologist with an interest in the past. The later chapters in C. E. P. Brooks *Climate Through the Ages* and a recent paper by H. W. Ahlmann, "The Present Climatic Fluctuation", *Geographical Journal*, CXII, 1948, suggest their general scope. The fruitful results that can be obtained by the investigation of the long runs of temperature figures that exist for some localities are demonstrated by Professor Manley's study of the "Temperature Trend in Lancashire 1753-1945", *The Quarterly Journal of the Meteorological Society*, Vol. LXXII, 1946. He shows that there is a consistent upward trend since 1830 in the January mean temperatures amounting to 3° F, though there is no corresponding change for June. An analysis of a long run of temperature figures for Stockholm shows a similar change; there is a rise of 4° F. in January mean temperatures since 1800. So our grandmothers are right after all, when, illustrating by tales of the delights of winter skating, they maintain that the weather has changed since their young days. The study of the climate of earlier periods will always be limited by lack of statistical data but the patient accumulation and correlation of indirect evidence may one day establish or disprove present surmises of increasing cold and wet in North-West Europe at the beginning of the present era, of amelioration here in the eleventh century, of increasing storminess in the North Atlantic in the fifteenth century, of drought whether of long or short periodicity in Central Asia. This knowledge will not only help towards a better appreciation of the geography of the Early Middle Ages, giving a better understanding of the widespread folk-movements of Goths and Franks, Anglo-Saxons and Jutes, Vikings and Danes, Slavs and Mongols, but it will also help the climatologist himself to see present conditions in perspective as the contemporary phase of an ever-changing climatic scene.

Whatever may be the advantage of the historical outlook to the physical geographer, the necessity of a historical approach to the biological, social, economic and political geographer cannot be denied. In some parts of the Amazon basin, in a few remote corners of Africa, in the dreariest wastes of the Atacama, the Australian and perhaps the Gobi deserts, in the northernmost frozen marshes of Arctic Canada and Siberia, there may be places where man, were he to settle, would write upon a clean sheet. Almost everywhere else present geographical features are etched in on top of many earlier designs. Sometimes old lines are etched deeper and deeper as they are repeated in each design; sometimes lines of an earlier design are re-etched and revived in a modern one, having been ignored and almost obliterated in intervening ones; sometimes new lines seem to be guided by the existence of the old approximately along old courses, or alternatively they are massed in the hitherto empty spaces; only rarely do new lines completely ignore the old, and when they do the design they make is often blurred by the old patterns underneath.

The grouping of living things, plants, animals and men within the area, the characteristic features of the group and of its distribution, depends in most cases upon a long association between society and place. The good heart of English soil in parts reflects the climate; but in part it reflects the long period of intensive working with relatively primitive tools, and the care of landlords in regulating the proportion of land that could be ploughed, the number of beasts that might be stocked, the amount of straw that could be sold off the farm so that a fat father should not make a lean son. The badly eroded soils of some areas of America reflect in part a difference in physical geography compared with Britain, but in part a difference in the time of their colonisation and therefore in the farm equipment and marketing opportunities of the settlers who exploited them. The tangled maquis of the more accessible slopes of the coastal mountains of Greece reflects not the demands of the modern economy but the work of man and his goats throughout many millennia. The political activities of man can never be separated from his social and economic activities and like them cannot be considered in the present without reference to the past. States in a modern sense are the creation of historic peoples; although some frontiers are highly unstable, many have been more or less fixed for long periods. It is astonishing how clearly old political frontiers show up in the landscape even if they are now no longer serving their original purpose: the lineaments of the political geography of Classical Greece and Rome still show on the modern map.

There is then a tangled skein of indissoluble ties that links the place as it is today to the place as it was in the past.

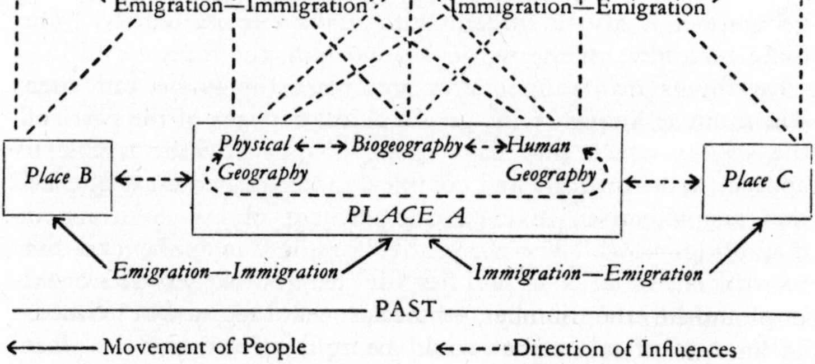

Present physical geography reflects not only the effects of the work of present imhabitants and of present societies, but the effects of the work of past inhabitants and of past societies. Present inhabitants and present societies do as they do in part in response to the opportunities that their present environment offers to them, but in part through tradition, in response to past environment, and perhaps to the past environment of another place. It is not sufficient to study the internal geography of an area in the past to appreciate the present; its external geography must be considered too. What society does in a place at any given time depends in part upon how much of the world it knows, and, of that which it knows, how much it uses, and of that which it uses, how it evaluates the resources.

Historical Geography forms then no ornamental coping to geographical study; it is with Physical and Biological Geography the foundation upon which the Social, Political and Economic Geography of the Modern World must rest. The *raison d'être* of work to establish the geographical patterns of the historic period, apart from its own intrinsic interest, lies not primarily in the fact that the geography of the past leaves a legacy in passive features of the landscape, impressive though many of these may be, but in the fact that in the conditions of the past is to be found so often the impetus that is still at work shaping and even initiating present geographical patterns.

The student of geography is a lucky one, he has his subject matter with him and about him all the time, has he but the eyes to see and the knowledge to interpret what he sees. But the geographer is also a student with grave responsibilities. The more he knows of the world around him, the more he realises what things man cannot alter and what things he can alter. The more he knows of the geographical patterns of the past and the way in which they have changed, the better he is able to appreciate the intricate and complexly inter-related consequences that any alteration in even one element of his environment may bring about. The more, therefore, is it incumbent on him to work, in so far as in him lies, for changes that would seem to make for the establishment, preservation and increase of harmony between man and nature to the lasting fruitfulness of the place.

BIBLIOGRAPHY

Chapter I

Dickinson, R. E. and Howarth, O. J. R.	*The Making of Geography* (1933).
Hartshorne, R.	*Perspective on the Nature of Geography* (1959).
ed. Chorley, R. J. and Haggett, P.	*Frontiers in Geographical Teaching* (1965).

Chapter II

Fisher, H. A. L.	*A History of Europe* (1936).
Trevelyan, G. M.	*History of England* (1934).
Trevelyan, G. M.	*English Social History* (1944).
Smith, C. T.	*Historical Geography of Western Europe* (1967).
Clapham, J. H. and Court, W. H. B.	*A Concise Economic History of Britain:* Vol. I, *from the earliest times to 1750* (1949); vol. II, *from 1750 to recent times* (1954).
ed. Darby, H. C.	*The Historical Geography of England before A.D. 1800* (1936).
Lynam, E.	*The Map Maker's Art* (1953).

Chapter III

Clark, Graham	*World Prehistory* (1961).
Clark, Graham	*Prehistoric England* (1940).
Hawkes, J. and C.	*Prehistoric Britain* (Pelican, 1944).

Charlesworth, M. P. and others — *The Heritage of Early Britain* (1952).

Fleure, H. J. — *A Natural History of Man in Britain* (1951).

Morant, G. M. — *The Races of Central Europe* (1939).

Huxley, J. S. and Haddon, A. C. — *We Europeans* (1935).

Adams, J. T. — *The History of the American People* (2 Vols., 1943).

CHAPTER IV

Ekwall, E. — *The Concise Oxford Dictionary of English Place Names* (4th edition, 1960).

Orwin, C. S. and C. S. — *The Open Fields* (in particular Part I.) (1938).

Orwin, C. S. — *A History of English Farming* (1949).

Franklin, T. B. — *A History of Scottish Farming* (1952).

Hoskins, W. G. and Finberg, H. P. R. — *Devonshire Studies* (1952).

Grigg, D. — *The Agricultural Revolution in South Lincolnshire* (1966).

Slicher van Bath, B. H. — *Agrarian History of Western Europe* (1963).

Brown, R. H. — *The Historical Geography of the United States* (New York, 1948).

Brown, H. C. — *Grandmother Brown's One Hundred Years, 1827–1927* (Boston, 1931).

Webb, W. P. — *The Great Plains* (1931).

CHAPTER V

Maitland, F. W. — *Township and Borough* (1898).

Tait, J. — *The Medieval English Borough* (1936).

Hill, J. W. F. *Medieval Lincoln* (1948); *Tudor and Stuart Lincoln* (1956).
Maxwell, C. *Dublin under the Georges* (1946).
Youngson, A. J. *The Making of Classical Edinburgh* (1966).
Stevenson, R. L. *Edinburgh* (1889).
Ormsby, H. *London on the Thames* (1924).
James, N. G. Brett *Growth of Stuart London* (1935).
Gilbert, E. W. *Brighton: Old Ocean's Bauble* (1954).
ed. Darby, H. C. *The Cambridge Region* (1938).
ed. Steers, J. A. *The Cambridge Region* (1965).
Dickinson, R. E. *The West European City* (1951).
Jones, E. *Towns and Cities* (1966).
Hiorns, F. R. *Town Building in History* (1956).

CHAPTER VI

Tansley, A. G. *The British Islands and their Vegetation* (1939).
ed. Darby, H. C. *Domesday Geography of England*, 6 vols. (1953–)
Albion, R. G. *Forests and Sea Power* (1926).
H.M.S.O. *Census of Woodlands* 1947–49 (Forestry Commission Census Report, No. I) (1952).
Darby, H. C. *The Medieval Fenland* (1940).
Darby, H. C. *The Draining of the Fenland* (1940).
Slater, G. *The English Peasantry and the Enclosure of Common Fields* (1907).
Gonner, E. C. K. *Common Land and Enclosure* (1912).
Emmison, F. G. *Types of Open-Field Parishes in the Midlands* (Hist. Assoc. Pamphlet, 1937).
Emmison, F. G. *Catalogue of Maps in the Essex Record Office* (1947).

Chapter VII

Salzman, L. F. — *Medieval English Industries* (1923).

Deane, P. — *The First Industrial Revolution* (1965).

Ashton, T. S. — *The Industrial Revolution 1760–1830* (Home University Library, 1948).

Hammond, J. L. and Barbara — *The Rise of Modern Industry* (5th Edn., 1937).

Usher, A. P. — *A History of Mechanical Inventions* (1929).

Dodd, A. H. — *The Industrial Revolution in North Wales* (1933).

Hamilton, H. — *The Industrial Revolution in Scotland* (1932).

Straker, E. — *Wealden Iron* (1931).

Court, W. H. B. — *The Rise of Midland Industries 1600–1838* (1938).

Ashton, T. S. — *Iron and Steel in the Industrial Revolution* (1924).

Ashton, T. S. and Sykes, J. — *The Coal Industry in the Eighteenth Century* (1959).

Nef, J. U. — *The Rise of the British Coal Industry* (2 Vols., London, 1932).

Lipson, E. — *History of the Woollen and Worsted Industries* (1921).

Heaton, H. — *The Yorkshire Woollen and Worsted Industries* (1920).

Ramsey, G. — *The Wiltshire Woollen Industry* (1943).

Mendenhall, T. C. — *The Shrewsbury Drapers and the Welsh Wool Trade in XVI and XVII Centuries* (1953).

Wadsworth, A. P. and Mann, Julia — *The Cotton Trade and Industrial Lancashire 1600–1780* (1931).

Daniels, G. W. — *The Early English Cotton Industry* (1920).

Gray, H. L. — "The Production and Exportation of English Woollens in the Fourteenth Century". *English Historical Review*, (Vol. XXXIX, 1924).

Carus-Wilson, E. M. *Medieval Merchant Venturers* (1954).
Redford, A. *Labour Migration in England 1800–1850.* (1926).
Wrigley, E. A. *Industrial Growth and Population Change* (1962).
Buer, M. C. *Health, Wealth and Population in the Early Days of the Industrial Revolution* (1926).
Cipolla, C. *The Economic History of World Population* (1962).

CHAPTER VIII

Harrison, William *Description of England*, 1577; 1587 (Ed., F. J. Furnivall 1877–1908).
Fiennes, Celia *The Journeys of Celia Fiennes, 1685–1703* (Ed., C. Morris, 1947).
Defoe, Daniel *A Tour thro' the Whole Island of Great Britain* (1724–27).
Young, Arthur *A Six Weeks' Tour thro' the Southern Counties of England and Wales* (1768).
Young, Arthur *A Six Months' Tour thro' Northern England* (1770).
Priestley, J. *Historical Account of the Navigable Rivers, Canals and Railways of Great Britain* (London, 1831).
Jackman, W. T. *The Development of Transportation in Modern England* (1916. Full Bibliography).
Willan, T. S. *River Navigation in England 1600–1750* (1936).
Willan, T. S. *The English Coasting Trade 1600–1750* (1938).
de Salis, H. R. *A Chronology of Inland Navigation in Great Britain* (1897).

de Salis, H. R.	*Bradshaw's Canals and Navigable Rivers of England and Wales* (1928).
Bradshaw, G.	*Railway Guides* (Edns. of various dates, First Edn., 1839).
Hadfield, C.	*British Canals* (1950).
Webb, S. and B.	*English Local Government: The Story of the King's Highway* (1913).

INDEX